SURGERY
and
THE ALLERGIC PATIENT

SURGERY
and
THE ALLERGIC PATIENT

Edited by

CLAUDE A. FRAZIER, M.D.

Asheville, North Carolina

With Forewords by

Warren H. Cole, M.D.

and

L. Nelson Bell, M.D.

CHARLES C THOMAS · PUBLISHER

Springfield · Illinois · U.S.A.

Published and Distributed Throughout the World by
CHARLES C THOMAS • PUBLISHER
Bannerstone House
301-327 East Lawrence Avenue, Springfield, Illinois, U.S.A.
Natchez Plantation House
735 North Atlantic Boulevard, Fort Lauderdale, Florida, U.S.A.

© *1971, by* CHARLES C THOMAS • PUBLISHER
Library of Congress Catalog Card Number 72-149182

With THOMAS BOOKS *careful attention is given to all details of
manufacturing and design. It is the Publisher's desire to present books
that are satisfactory as to their physical qualities and artistic possibilities and appropriate for their particular use.* THOMAS BOOKS *will
be true to those laws of quality that assure a good name and good will.*

Printed in the United States of America
N-1

Contributors

Claude A. Frazier, M.D., *Editor*
Asheville, North Carolina

John Adriani, M.D.
Director, Department of Anesthesiology
Charity Hospital of New Orleans
Professor of Surgery
Tulane University School of Medicine
New Orleans, Louisiana

Brian Blades, M.D.
Chairman, Department of Surgery
Lewis Saltz Professor of Surgery
George Washington University School of Medicine
Washington, D.C.

Alexander A. Fisher, M.D.
Clinical Professor, Department of Dermatology
New York University Post-Graduate Medical School
Associate Attending in Dermatology
University Hospital New York University Medical Center
New York, New York

Claude A. Frazier, M.D.
Chief of Allergy
Memorial Mission Hospital
and St. Joseph's Hospital
Asheville, North Carolina
Consulting Editor, Review of Allergy

Roy F. Goddard, M.D.
Director, Pediatric Research Department
Lovelace Foundation for Medical Education and Research
Medical Director, Pediatric Pulmonary Center of New Mexico
Regional Medical Program
Medical Director, New Mexico Cystic Fibrosis Center
Clinical Associate, Department of Pediatrics
University of New Mexico Medical School
Albuquerque, New Mexico

Robert L. Goodale, M.D.
Aural Surgeon (Retired)
Senior Consulting Surgeon
Massachusetts Eye and Ear Infirmary, Boston, Mass.
Ipswich, Massachusetts

Harold A. Lyons, M.D.
Professor of Medicine, State University of New York
Downstate Medical Center
Director, Pulmonary Disease Division and Cardiopulmonary Laboratory
State University—Kings County Medical Center
Brooklyn, New York

Thomas L. Marchioro, M.D.
Professor of Surgery, Markle Scholar in Medicine
University of Washington School of Medicine
Seattle, Washington

William R. Richardson, M.D.
Assistant Clinical Professor of Surgery
University of Cincinnati College of Medicine
Attending Surgeon
Children's Hospital and Good Samaritan Hospital
Cincinnati, Ohio

William B. Sherman, M.D.
Associate Clinical Professor of Medicine
College of Physicians and Surgeons
Columbia University
Director of the Institute of Allergy
Roosevelt Hospital
New York, New York

James E. Shira, M.D., Lt Colonel, MC, U.S. Army
Assistant Chief, Pediatric Service
Chief, Pediatric Allergy Section, Fitzsimons General Hospital
Assistant Clinical Professor of Pediatrics
University of Colorado Medical Center
Denver, Colorado

Frederic Speer, M.D.
Associate Clinical Professor of Pediatrics
University of Kansas and University of Missouri
Chief of Pediatric Allergy, University of Kansas Medical Center and Children's
   Mercy Hospital
Kansas City, Kansas

To my father, Dr. Claude Frazier of Huntington, West Virginia — a physician whose life and works have made an indelible impression upon my life.

Foreword

ALLERGY APPEARS to be increasing, perhaps because of the increasing use of drugs and also because many of the allergic reactions were not recognized as such years ago. Allergic complications are particularly important to the surgeon because they may occur during or immediately after the operation and threaten life.

Dr. Frazier's recognition of the importance of allergy to the surgeon has prompted him to compile this book, which is particularly significant since very little is found in the medical literature dealing with the numerous examples of allergy in surgical patients. Chapter 1 deals with anaphylaxis; it is very appropriate that the serious aspects of allergy be presented first and that this chapter be written by an expert in allergy; namely, the editor. Dr. Frazier describes the problems of allergy to the numerous drugs and emphasizes the various precautions in prevention, as well as the emergency treatment, of allergic reactions.

The editor has divided the material into eleven chapters, which are extremely well chosen. Since the allergies in children are so different from those in adults, separate chapters very appropriately have been written for these two groups. In fact, three chapters deal specifically with children and three with adults.

The editor is to be congratulated on his choice of authors for the various chapters. With no exceptions he has chosen authors who are particularly well informed with the material in their respective chapters. As stated in Chapter 8, allergy to drugs is especially important in surgery because so often the reaction occurs during or immediately after the operation. It is very fitting, indeed, that this chapter be written by a very experienced anesthesiologist who is likewise an expert pharmacologist. One could not find an author more capable than Dr. Adriani to write this chapter because his experience with anesthesia, drugs and pharmacology is practically unmatched anywhere.

Another chapter very appropriately chosen is the one on organ transplantation. Although this chapter is splendid, we must admit the important features of immunology in transplantation are not yet known. The present knowledge of organ transplant immunology allows us to safely transplant kidneys, but not livers, heart, lungs and other organs. We need better drugs for these organs. We do not know when they will be available, but we do know that modern science does not stand still. It is only a matter of time when practically all organs can be transplanted safely.

This book is very important to the surgeon; he should be familiar with its contents. Needless to say most of the material in it is applicable to the medical aspects of allergy. Therefore, the volume will be very useful to all specialists as well as students.

Warren H. Cole

Foreword

SURGEONS FOR many years were inclined to regard allergies and allergic reactions as a purely medical phenomenon unrelated to their own specialty
and therefore of no more than passing interest to them and their patients.

Some tragic and near-tragic experiences have changed all that and it is
now imperative that the surgeon know something of the nature, occurrence
and treatment of allergic episodes because he is himself confronted with
them in his patients.

It is probable that the complexity of modern life has brought with it
more different forms of allergy than those recognized in the past, with more
people now being subject to their accompanying reactions. For this reason
Dr. Frazier and those with whom he has associated himself in this book
are to be congratulated on the depth of scientific knowledge made available to the surgeon and are to be thanked for the insights and practical aid
which they offer to both surgeon and patient.

This book is unique and tremendously important, and every surgeon
should fortify himself with the information so effectively presented.

L. Nelson Bell

Preface

ABOUT 25 percent of the population of the United States is presently suffering from some allergic condition. The surgeon comes in contact with a large number of these patients in his practice. In this book we will attempt to help the surgeon prevent and treat these allergic conditions and, if need be, to point out when to refer the patient for further definitive treatment.

Allergic conditions may be present when the surgeon first examines the patient. However, sometimes an allergy will develop as a result of medications given or procedures performed in the hospital or in the surgeon's office. Often a previous allergy, such as asthma or urticaria, may flare up while the patient is under the surgeon's care. When this occurs, the surgeon must know what to do to relieve the symptoms or what at least to give for emergency treatment.

Our book, written primarily for the surgeon, will familiarize him with the various allergies he may encounter in his surgical practice and will present information on the prevention, diagnosis and treatment of these allergies.

Acknowledgments

I AM GRATEFUL to Miss Martha E. Dana, Medical Editor at Hoffman-La-Roche, Inc., Nutley, New Jersey, for her assistance in editing this book.

As it was my good fortune to obtain such outstanding contributors for this volume, it is with pleasure that I acknowledge these men for their time, work and cooperation in the compilation of this material.

I thank Dr. Warren H. Cole for reading the manuscript and offering some very valuable suggestions.

Gratefulness is extended to all persons whose encouragement, work and comments helped to make this book become a reality. To Charles C Thomas, Publisher, a special thank you for the skill and efforts put forth to make this book an important contribution to medical literature. I would also like to thank my secretary, Miss Sherry A. Morrow, for her conscientious efforts in all the clerical work. Finally, I would like to express appreciation to my wife, Kay, for her painstaking effort in reviewing the final draft of this manuscript.

Contents

SURGERY

and

THE ALLERGIC PATIENT

Chapter 1

Anaphylaxis: Prevention and Treatment

Claude A. Frazier

EVERY PHYSICIAN who gives or is responsible for giving sensitivity-producing drugs must have some basic knowledge of drug allergy. He must know which of the drugs he prescribes or which his patient takes on his own responsibility can produce sensitivity. He must know all about his patient's past history of any allergy, previous exposure to allergenic drugs, previous unfavorable reactions from drugs, and family history concerning allergy. As a precautionary measure, I ask all my patients to bring me a list of drugs they have taken and those presently being taken; and I request that they also record the names of any new drugs that are prescribed. Previous exposure to a drug (except in rare instances of adverse reaction on first exposure) usually sets the stage for drug sensitivity.

A direct relationship has been suggested between the number of drugs a patient receives and the probability of his acquiring an adverse reaction.[1] A review of the patients in the medical service of one hospital revealed that over a four-week period, 15 percent experienced drug reactions. Each one of a group of thirty-eight patients with reactions to a specific drug had received an average of fourteen different drugs during hospitalization.

Unknown exposure must also be given consideration, for example, that resulting from penicillin in milk or in certain vaccines and from silk used in the filtration of certain vaccines during their preparation.

A history of previous exposure to a drug without untoward effect, however, does not rule out present sensitivity to the same drug. Moreover, cross-reactions tend to occur in patients sensitive to one drug when they are given other agents from within the same chemical group.

The symptoms of anaphylaxis include flushing, pallor, palpitation, cyanosis, nausea, vomiting, abdominal cramps, diarrhea, nasal congestion, sneezing, red and watery eyes, coughing, itching, urticaria, angio-edema, dyspnea, frothy sputum, wheezing, fainting, shock and convulsions.

PREVENTION OF ANAPHYLAXIS

Prevention is the most effective form of treatment for anaphylaxis. The following are some basic rules for the prevention of such allergic reactions to drugs:

1. *Take a history.* Before prescribing any drug, take a careful history of the patient's drug sensitivity.

2. *Do believe the patient who says he is allergic to a drug.* Write in red on the outside and the inside of the patient's chart that he is allergic to a certain drug. It is imperative that this be recorded when first learned from the patient. Do not run the risk of simply recording the drug sensitivity somewhere within the chart; this can easily be forgotten and overlooked at later dates. In addition, tell your patient to remind you—and every physician he consults—of his drug sensitivity. Register the patient with the Medic Alert International Foundation and have him wear the identification tag identifying his allergy.

3. *Substitute unrelated agents for allergenic drugs.* If there is a positive history of allergy to the drug proposed for use, another unrelated therapeutic or diagnostic agent should be employed. This includes using acetaminophen (Tylenol®) for acetylsalicylic acid (aspirin), lidocaine for procaine and hyperimmune human tetanus antitoxin instead of heterologous antitoxin.[2]

4. *Do avoid unnecessary sensitization.* By far the most common cause of unnecessary sensitization is the routine use of antibiotics (especially penicillin) for the most trifling infections. More than half of all the prescriptions written in the United States are for antibiotics; more than a third of the nation's drug bill is for antibiotics. Much indiscriminate use exists and is to be deplored.

It is important to remember that reactions may occur as a result of both topical and systemic use. Topical application is most likely to sensitize, parenteral administration ranks next most likely, and oral administration the least likely, therefore the safest. Furthermore, topical administration of an antibiotic may sensitize the individual who later may need the drug for severe infection. Therefore it is better to avoid the use of topical penicillin and tetracycline since there are other topical antibiotics in ointments that may be used, such as gramicidin, bacitracin, oxacillin and polymyxin.

5. *Do not overlook indirect sources of antibiotics.* When patients known to be sensitive to antibiotics continue to show adverse reactions though these drugs have not been prescribed, exposure from indirect sources must be considered. One possibility is that other pharmaceutical products prescribed were contaminated with an antibiotic during manufacture. The Federal Drug Administration has recently ordered manufacturers to institute protective measures to insure other products remain pure.

Another obscure source of unwanted penicillin has been milk. Also, syringes used previously for penicillin injections, despite sterilization, sometimes contain sufficient residual amounts of antibiotic to precipitate serious reactions in highly sensitive patients, and even sterilizers have occasionally become contaminated from such sources.

6. *Observe the patient after injection of a drug.* Whenever possible, after the first injection of a drug, keep the patient in the office under observation for fifteen to twenty minutes.

7. *Remember that skin tests to most drugs are unreliable.* It is also possible for a serious reaction to develop from a skin test. If a test is to be performed, a scratch test should be done initially. A negative skin test does not guarantee that there will be no systemic reaction when the drug is given parenterally.

If a person is to receive horse serum, skin tests are most helpful.

The initial scratch test is followed by an intradermal test with 0.02 ml of 1:100 dilution.[3] Persons who experience asthma when exposed to horses usually react adversely to horse serum.

If one plans to use the tetanus antiserum, a safer alternative is to use human antitoxin. Previous immunization with tetanus toxoid would forestall this problem.

If there is no contraindication, injection of a new therapeutic agent should be made in an upper extremity so that a tourniquet can be applied and epinephrine injected locally if a reaction should occur.

8. *Know anaphylaxis-producing drugs.* One should be familiar with agents that may produce anaphylactic reactions in man. Some of these are horse antiserum; hormones such as relaxin, insulin, corticotropin; enzymes such as chymotrypsin, trypsin, penicillinase; and dextran, thiamine, sulfobromophthalein, procaine, salicylates and antibiotics, especially penicillin.[4]

9. *Do not be complacent about prescribing aspirin.* In the past decade, consumption of aspirin has doubled in the United States. Intolerance to aspirin is not uncommon in individuals who are near middle age and who, as a rule, do not have a history of atopy. It is characterized by changes in the skin and respiratory mucous membranes—angio-edema, perennial rhinitis, formation of nasal polyps, and bronchial asthma. Patients who have an intolerance to aspirin have a comparable intolerance to other analgesics.[5]

A large number of patients with the combination of chronic paranasal sinus disease, nasal polyp, bronchial asthma and chronic pulmonary emphysema are unusually sensitive to aspirin. Swelling of the mucous membranes of the mouth, larynx and bronchi may follow instantaneously after its use. Aspirin should be avoided in such patients.

10. *Do use particular caution with penicillin.* Of all the antibiotics, penicillin produces the greatest number of serious reactions and consequently is the most extensively investigated with respect to immunologic mechanisms of adverse effects. Each year in the United States an estimated one hundred to three hundred fatalities are caused by penicillin.[6] Most deaths have occurred without warning in persons having no history of allergy.

An expert committee of the World Health Organization[7] recommended in 1959 that individual and public health measures should be taken to prevent or treat penicillin reactions with particular reference to anaphylactic reactions. These measures are summarized below:

1. Always have an emergency kit for treatment of allergic reactions readily available.

2. Always have an exact history of the patient's previous contact with penicillin, previous penicillin reactions and allergic diathesis.

3. No penicillin treatment should be given to patients with a previous history of reaction.

4. If possible, refer patients with suspected penicillin allergy to a specialist trained in modern immunologic techniques.

5. Always tell the patient that he is going to receive penicillin treatment.

6. No penicillin should be employed topically either externally or on mucous membranes.

7. Avoid the use of penicillinase-resistant penicillins (methicillin, cloxacillin, nafcillin, ancillin and quinacillin), which should be reserved for infections caused by penicillinase-producing staphylococci.

8. Ensure the thorough washing and adequate sterilization of all-purpose syringes used in penicillin treatments when re-using them to inject other drugs.

9. If possible, detain all patients for half an hour in the clinic after an injection of penicillin.[7]

Skin-testing for wheal and flare reactions with penicillin and penicilloyl-polylysine prior to therapy is the most practical and efficacious means known presently for the detection of potential anaphylactic reactions. At present, however, intradermal skin tests with PPL or penicillin should be performed only by qualified clinical investigators because anaphylactic reactions have been reported with even minute skin test doses.[8]

If a therapeutic agent must be used despite a positive history of allergy, precautionary measures should include starting an intravenous infusion to facilitate the prompt administration of epinephrine, diphenhydramine hydrochloride and other drugs to treat anaphylaxis and vascular collapse.[4] A tracheostomy set should also be available.

The symptoms of anaphylaxis may be mild or severe, but their progression to death can be very rapid. The physician must be alert for the earliest symptoms, and he should train his office staff to recognize and report them. Seconds are important. I recall that early in my practice one of my patients began to rub her eyes about fifteen minutes after having received an injection. My secretary said the patient felt there was something in her eye. In the treatment room she began to sneeze, her eyes were red and watery and a few urticarial wheals appeared. With immediate treatment for anaphylactic reaction, the symptoms subsided quickly. Since that time I have routinely trained all my staff, secretaries as well as nurses, in the early recognition of the symptoms of anaphylaxis.

EMERGENCY KIT

An emergency kit should be prepared with all drugs and with equipment together and available at a moment's notice. I have kept one in my office since opening practice. The following drugs and equipment are suggested:

Drugs

Epinephrine HCl solution (1:1000) 1 cc ampules, ×5
Aminophylline (injectable) 250 mg/10 cc, 2 vials
Injectable antihistamine (diphenylhydramine HCl 50 mg/cc, 2 ampules)
Intravenous steroids such as hydrocortisone sodium succinate (Solu-Cortef®) or prednisone phosphate (40 mg/2 cc vial)
Injectable vasopressor (Aramine® or Levophed®)
5% glucose in saline for intravenous infusion
Amytal® for injection
Neo-Synephrine® 1 mg/cc ampules, ×3.

Optional Drugs

Chlorpromazine	Phentolamine methanesulfonate
Glyceryl trinitrate tablets	Isoproterenol HCl
Digoxin	Potassium chloride
Cedilanid®	Sodium lactate injection
Deslanoside injection	Sodium bicarbonate injection
Quinidine gluconate	Diphenylhydantoin
Atropine sulfate	Calcium gluconate

Equipment

Alcohol sponges	Suture material
Sterile syringes and needles	Endotracheal tube and oral airway sizes
Adhesive tape	
Tourniquets	Hand-operated respirator
Scalpel	Aspirator
Hemostats	Oxygen should be nearby

Hospital Emergency Equipment (Additional Items)

Tracheostomy box	Phlebotomy tray, suture boat, intra-catheters
Ventilator	
Nasal oxygen unit and catheters	Blood pump
Cardiac board	Electrocardiograph
	Defibrillator

TREATMENT

At the first sign of anaphylaxis, it is important to begin treatment. Epinephrine 1:1000 should be injected subcutaneously, 0.1 to 0.2 cc in the very young child; 0.2 to 0.3 cc in the older child and 0.3 to 0.5 cc in the adult. It is important to realize that the dosage must be individualized. If the patient responds to epinephrine, he must be observed carefully every few minutes as the effect of the initial dose wears off. Epinephrine may be repeated every one to twenty minutes if necessary.

If the anaphylaxis occurred from an injection into an upper extremity, a tourniquet should be placed proximal to the site. Injection of 0.1 to 0.2 cc epinephrine 1:1000 in the injection site may help to delay absorption of the anaphylactic agent.

If bronchospasm develops despite the administration of epinephrine, aminophylline should be administered slowly intravenously in a dosage of 0.25 to 0.5 gm in adults. Most children cannot tolerate more than a dose of 3 mg per kilogram of body weight ($\frac{1}{10}$ grain per pound) .

An antihistamine should be injected intramuscularly or intravenously. Diphenhydramine (Benadryl®) can be given intravenously in a dosage of 50 mg for adults and older children, 25 mg for young children six to twelve years of age. For the child from one to six years old, it can be given intravenously in a dose of 1 mg/kg, not to exceed 50 mg.

If it is necessary to hold and prolong the effects initiated by these measures, steroids by intravenous drip are indicated. While I prefer Solu-Cortef, 100 mg, injected intravenously, prednisone phosphate, 40 to 80 mg, intravenously, may be given instead. Corticosteroids do have a beneficial effect but should be used only to supplement other measures. They are not to take the place of epinephrine.

The steroids may be repeated every six to eight hours as needed. Since one is dealing with an acute process, generally not more than thirty-six hours of corticosteroids is needed. Oxygen will minimize the development of hypoxia.

If there has been a marked drop in the blood pressure, the patient should be placed in shock position. Aramine® injected intravenously is suggested. For a child under twenty pounds of weight a dose of 0.5 mg is suggested; for the older children, 1 mg. For transient circulatory collapse phenylephrine hydrochloride (Neo-Synephrine) can be given.[3] Intramuscular administration of 5 mg results in elevation of the blood pressure for one to two hours; 0.5 mg intravenously produces a quick effect that lasts only ten to twenty minutes; or 10 to 20 mg may be added to a liter of intravenous solution and the rate of administration adjusted on the basis of blood pressure response. Occasionally it may be necessary to replace blood volume with plasma.

Laryngeal edema may occur with upper airway obstruction. This may necessitate an oral airway or an endotracheal tube in unconscious patients. Tracheostomy and tracheotomy may be required.

In case of cardiorespiratory arrest, cardiac massage and artificial respiration is necessary. Therapy within three to four minutes is essential to avoid permanent brain damage.[9]

Expert assistance should be obtained immediately—from the hospital's cardiac arrest team if one is available. ABCD steps to remember[10] are

> A. . . . airway opened
> B. . . . breathing restored
> C. . . . circulation restored
> D. . . . definitive therapy.

If no chest wall or abdominal motion can be discerned, not enough air is being moved. In checking the cardiac status, one feels for the peripheral pulse and listens over the precordium. If the patient is apneic and there is no heartbeat, artificial respiration and cardiac resuscitation must be attempted.

One should be certain that the airway is not obstructed by the tongue, vomitus, false teeth, or other foreign bodies and that it is maintained. False teeth or other foreign bodies must be removed. The neck should be hyperextended by grasping the angle of the jaw. A nasal or oral airway may relieve obstruction. If one is skilled in intubation, an endotracheal tube is even better. A tracheostomy may be necessary if all else fails.

Once the airway is opened, respiration can be given mouth to mouth or by bag and mask if available with oxygen (Fig. 1-1). The rate of artificial respiration should be 15 to 20 per minute. One should guard against overdistending the very young patient's lungs. One guide is to observe the rise and fall of the chest. A newborn infant may require only a puff of air.

External cardiac resuscitation is most effective when the patient is on a firm surface. The operator, at arm's length over the patient's chest, places the heel of one hand over the xiphisternum and the other on top of it. He applies pressure by leaning down with the weight of his body. The patient's chest may be depressed $1\frac{1}{2}$ to 2 inches; 80 to 120 pounds of force is used. (The technique can be practiced by using a bathroom-type scale [Fig. 1-2]). Be careful to keep fingers off the chest wall; the cartilaginous ends of the ribs may yield.

Effectiveness may be improved by a prolonged sternal compression of 0.5 second rather than a quick thrust. The rate of compression should be 60 per minute in an adult; for infants and children 60 to 90 per minute.

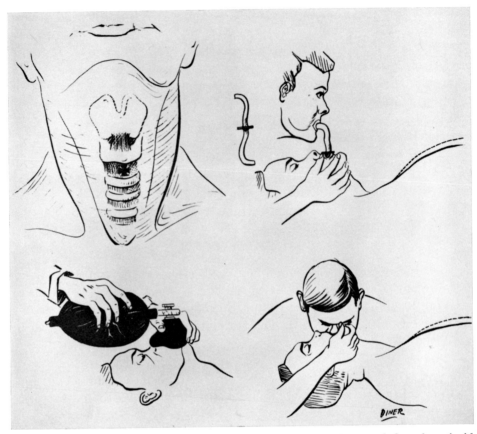

Figure 1-1. Pulmonary resuscitation. (*Upper left*) "X" marks the spot below the cricoid where a 15-gauge needle may be inserted for a quick tracheostomy. (*Upper right*) Convenient resuscitation with double-ended airway which keeps two sizes available in a single device. The patient's lips should be sealed against the airway. Rise and fall of the chest wall indicted effective air movement. (*Lower right*) Without an airway the chin must be pulled forward and the neck vigorously hyperextended to free the tongue from obstructing the air passage. Note that the resuscitator's cheek occludes the victim's nostrils. (*Lower left*) One type of hand-operated respirator. The mask must be pressed firmly against the face. If an airway is not inserted, the position of the head and neck must be the same as that in mouth-to-mouth resuscitation. (Courtesy of Radiological Clinics of North America, Barnhard and Barnhard.[9])

In a child one hand may be used for external cardiac compression. In younger children a few fingers may be used, and in an infant one's fingers can be interlaced behind the infant's back and thumbs superimposed over the sternum. The younger the child, the higher the heart is in thorax so that in an infant pressure should be applied to the midsternum.

Effective cardiac compression will result in improved color, constricted

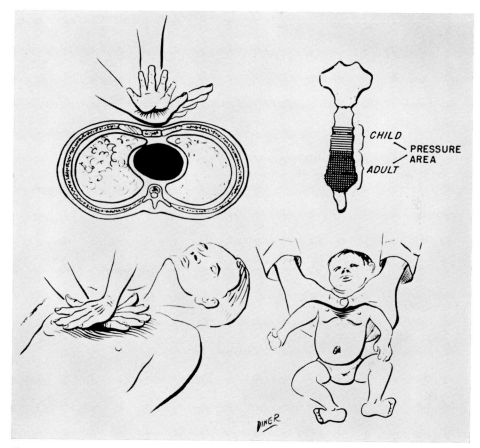

Figure 1-2. External cardiac compression. (*Left*) Placement of hands on adult chest so the sternum is pressed against the heart without simultaneous compression of the ribs. (*Lower right*) The fingers are interlaced behind the back of a child while superimposed thumbs transmit pressure to the sternum. (*Upper right*) Pressure area in the adult is just above the xiphisternum while that in a child is at the midsternum. (Courtesy of Radiological Clinics of North America, Barnhard and Barnhard.[9])

pupils, spontaneous gasping, swallowing and recordable blood pressure with each compression.

If the physician is alone, he can start mouth-to-mouth resuscitation a few times and then alternate three to four cardiac compressions for each artificial respiration. If external compression fails, a qualified individual should perform open cardiac massage.

Acidosis may result from anoxia. If this occurs, sodium bicarbonate should be given intravenously, 3.75 gm, (44.6 mEq) or more every five minutes.[11] If the heartbeat is restored and the blood pressure is over 70 mm Hg, Aramine and Levophed can be used to maintain blood pressure. Aramine, 5 to 10 cc of a 1% solution (10 cc vial) in 500 cc saline or 5% dextrose in water, or Levophed, 4 to 8 cc of a 0.2% solution (4 cc ampules) in 500 cc saline or 5% dextrose in water, may be given to maintain the blood pressure.

If the heartbeat is not restored in a few minutes, an electrocardiograph is indicated. If the heart is atonic, adrenalin, 0.2 cc (1:1000), or isoproterenol hydrochloride, 0.1 cc (200µg/cc) in 10 cc saline, can be injected into the left ventricle through a long spinal needle. The injection is made through the fourth left intercostal space at the midclavicular line. If ventricular fibrillation is suspected, an external defibrillator can be used.

The management of anaphylaxis includes a knowledge of drug allergy, recognition of the symptoms and prompt institution of therapy.

REFERENCES

1. Cluff, L.E., Thornton, C.F., and Seidl, L.G.: Studies on the epidemiology of adverse drug reactions. I. Methods of surveillance, *JAMA, 188:*976, 1964.
2. Sheffer, A.L.: Therapy of anaphylaxis, *New Eng J Med, 275:*1059, 1966.
3. Henderson, L.L.: Anaphylaxis, *Ann Allerg, 23:*528, 1965.
4. Austen, K.F.: Systemic anaphylaxis in man. *JAMA, 192:*116, 1965.
5. Samter, M.: Intolerance to aspirin. *Ann Int Med, 68:*975, 1968.
6. Northington, J.M.: Penicillin hypersensitivity. *Clin Med, 71:*803, 1964.
7. Idsoe, O., Guthe, T., Willcox, R.R., and DeWeck, A.L.: Nature and extent of penicillin side-reactions, with particular reference to fatalities from anaphylactic shock. *Bull WHO, 38:*181, 1968.
8. Hewitt, W.L.: The penicillins: A review of strategy and tactics, *JAMA, 185:*264, 1963.
9. Barnhard, H.J., and Fay, M.: Emergency treatment of reactions to contrast media: Updated 1968. *Radiology, 91:*74, 1968.
10. National Academy of Sciences-National Research Council: Cardiopulmonary resuscitation. Statement by the Ad Hoc Committee on Cardiopulmonary Resuscitation of the Division of Medical Sciences. *JAMA, 198:*372, 1966.
11. Orkin, L.R.: Comment on Dr. H.R.S. Harley's abstract, Reflections on Cardiopulmonary Resuscitation. *Survey of Anesthesiol, 11:*340, 1967.

Chapter 2

Medical Management of the Allergic Adult Before and After Surgery

William B. Sherman

IT IS ESSENTIAL that the surgeon undertaking operations on allergic patients be familiar with the essential nature of allergic diseases. Briefly, since this subject is discussed in detail elsewhere,[1,2] the allergic patient is so sensitized to one or more foreign substances, generally called antigens or allergens, that contact with them causes the local or general disturbances of body function designated allergic disease. The sensitization is the reaction of the body's immune mechanism to previous contact with the antigen, and it has the high degree of specificity characteristic of reactions of the immune mechanism. Thus the sensitized individual at first reacts only to the original antigen and those substances closely related to it in chemical structure. This basic sensitization usually lasts many months or years and later may be complicated by additional reactions.

Symptoms of allergic disease vary with the degree of exposure to the antigen and are often intermittent, or they are seasonal when the offending antigens such as pollens and molds occur seasonally.

Many symptoms and diseases affecting various organs and tissues have been attributed, with greater or less scientific evidence to allergy. Asthma, hay fever and related forms of rhinitis, urticaria, infantile eczema and contact dermatitis are among the most familiar and best established diseases of allergy.

Even the briefest discussion of allergy must distinguish between two general types of allergic reactions: (a) the delayed reaction in which sensitized cells react directly with the foreign material (contact dermatitis) and (b) the immediate reaction in which the interaction of sensitized cells and the allergen causes release of active chemical mediators into tissue fluid and circulation which have a profound effect upon the physiologic activities of susceptible tissues and organs, chiefly the small blood vessels, smooth muscles and mucous glands (hay fever). Among the half-dozen different substances suspected of acting as chemical mediators of immediate reactions,[3] histamine is the most familiar and probably one of the most important.

The hypersensitivities in both the immediate and delayed types of reactions may develop as the result of normal functions of the immune mechanism closely related to the development of resistance to infective agents. Both can readily be induced in experimental animals and in essentially all human subjects by sufficient exposure to suitable allergenic foreign agents.

However, in a large proportion of the cases of clinically important allergic diseases of the immediate type of hypersensitivity, an hereditary atopic factor renders some 10 percent of the population abnormally susceptible to sensitization. In such patients, apparently because of an abnormal immune mechanism, sensitization develops as a result of casual exposures to antigens which have no effect upon other persons. This atopic hypersensitivity is the basis of many of the commonest and most clinically important allergic diseases such as infantile eczema, hay fever and asthma. Since the tendency to their development is hereditary, they particularly affect certain individuals and certain families. Such individuals are frequently allergic to a wide variety of foods, inhaled antigens and other substances with which they have only the normal contacts.

It should be noted that, while the underlying hypersensitivity has the high degree of specificity typical of immune reactions, the hypersensitivity of chronic allergic diseases is less specific. After the reactive organ has been repeatedly stimulated by the specific antigen, it becomes so irritated and irritable that contact with a wide variety of nonspecific agents may produce essentially the same reaction as the original antigen. Thus, the patient with strictly seasonal bronchial asthma due to a specific pollen may later have attacks in other seasons precipitated by respiratory infections, nonspecific chemical irritants such as fumes and smoke, and by respiratory exertion, sudden changes of temperature and probably also by purely emotional factors. The influence of such nonspecific secondary factors greatly complicates the clinical management of allergic diseases.

FACTORS INFLUENCING SURGICAL PROCEDURES

The presence of specific sensitization and of allergic diseases due to it may be an important factor in surgical treatment of the patient for a number of reasons. The existence of sensitization presents the possibility of allergic reactions to agents used in anesthesia and surgery and in the earlier diagnostic procedures that involved the injection of foreign material, such as contrast media for x-ray films. The actual occurrence of active allergic disease further complicates surgical procedures. For example, ventilatory impairment due to bronchial asthma may make general anesthesia difficult and greatly increase the possibility of postoperative complications; a nasal mucosa that is inflamed and edematous because of allergy may increase infection in a nose and throat operation and thus delay healing; widespread or generalized allergic skin rashes may make sterilization of the skin difficult and greatly increase troublesome local complications in any form of surgery.

Finally, the widespread use of corticosteroid drugs, powerful and useful agents for the treatment of allergic disease, has introduced new problems into surgery upon patients so treated. Most of the older drugs used in the treatment of allergic disease—the adrenergic amines, theophylline and its derivatives, and the antihistamines—have relatively brief periods of action. They may temporarily disturb the normal physiology of the body but their effects last only a day or two. The action of the corticosteroid drugs, on the other hand, depends upon creation of a state of induced hypercorticoidism which profoundly disturbs the endocrine balance of the body; recovery of normal functions is prolonged over weeks and months. In many cases, this artificial disturbance of the endocrine balance suppresses normal functions of the adrenal cortex which, if prolonged over a sufficient period of time, may lead to atrophy. Normal adrenal cortical activity is important in the adaptation of the body to serious stress, therefore its depression or atrophy may cause surprising and disastrous failure of the normal adaptation and tolerance for surgery. The reactions of patients treated with the same type and amount of corticosteroids vary greatly, and the speed of the recovery is uncertain. The possibility of such reactions is serious in any patient who has received one of the corticosteroid drugs for several weeks in the few months previous to surgery.

PREOPERATIVE STUDIES

Both the nature and severity of the patient's allergic disease and the type and urgency of the proposed surgical treatment determine the extent of preoperative medical care of a patient with allergy. In the case of elective operations, a prolonged period of study of the allergic disease and numerous preoperative procedures are desirable and usually possible. The same procedures may be quite impossible when an emergency operation is necessary. However, even in the most urgent case, a quick evaluation of the nature and severity of the allergic state and of the drugs, particularly corticosteroids, used for its treatment is essential.

Adequate discussion of the preoperative study and care must distinguish among (a) elective operations and urgent operations for concurrent conditions not necessarily related to the allergic state, (b) operations on the tissues or organs which are the site of active allergic reactions, (c) operations intended for the amelioration of the allergic disease. The latter operations are discussed in other chapters of this book.

Preoperative Control of Allergy Before Elective Surgery
Allergies with Concurrent Disease

Elective surgery for concurrent disease in the allergic patient offers the greatest opportunity for evaluation and control of the allergic state before operation. The relative severity and course of the allergic disease and that of the disease or condition for which surgery is proposed determine the preoperative time available for treatment.

If the allergic disease is of major importance, the allergic state should be carefully investigated and evaluated.[2] Such a survey should determine, by a review of previous records or by a new history and skin tests, the relative importance of seasonal factors such as pollen and molds; of environmental contacts in the home, at work and in other activities; the effect of foods and drugs ingested; and the presence of infective factors which may influence the course of the allergic disease.

If the history indicates that the allergic activity has had important seasonal variations, the date of operation should be planned for a season when the patient has been relatively free of allergic symptoms in previous years. If the allergic symptoms prove to be largely due to nonseasonal environmental contacts, considerable relief may be expected when contact is avoided by hospitalization for four or five days. In such cases, it is important to allow a period of rest in a clean, preferably air-conditioned, room for several days before the operation. This period also offers opportunity for the treatment of the allergic disease by medications and other symptomatic measures before the operation is undertaken.

Asthma

If the patient for whom surgery is planned suffers from bronchial asthma, effective preoperative control of the disease is essential to permit satisfactory anesthesia and to lessen the incidence of postoperative pulmonary complications. Asthma may be caused by exposure to a specific extrinsic allergen or by respiratory infection, or a combination of these two factors.

If an extrinsic allergen is implicated, asthma may be controlled by avoiding exposure to it, by desensitization to lessen the degree of sensitivity, or by drugs which block or ameliorate the reaction. Reduction of exposure to an environmental allergen, accomplished by several days of hospitalization in a clean, air-conditioned room, usually produces partial relief. Since desensitization requires weeks or months of treatment, it is rarely feasible as a preoperative procedure. Drug treatment usually produces more rapid results and is, therefore, the most reliable treatment when time is an important factor.

In the treatment of asthma, the most effective drugs fall into three classes: (a) the adrenergic amines such as epinephrine isoproterenol and ephedrine; (b) theophylline and its derivatives; and (c) the adrenal corticosteroids. Other drugs may be used. The antihistamines, when used alone, are relatively ineffective but are often useful as adjuncts to treatment.

Iodides and other expectorants are helpful when there is thick, tenacious mucoid sputum. Antibiotics are important for the control of concurrent or secondary respiratory infections.

The choice of drugs depends upon the stage and severity of the asthma. For control of the acute attack, the subcutaneous injection of epinephrine (1:1000, 0.3 to 0.5 ml), repeated if necessary, usually produces quick results. A more prolonged effect may be attained by the use of more slowly absorbed preparations in oil or aqueous suspension. For recurrent attacks, self-medication by inhaling a spray of isoproterenol or epinephrine, or the use of oral ephedrine is convenient. Many proprietary combinations of ephedrine, theophylline and a mild sedative are available and useful.

Attacks not relieved by ephedrine or epinephrine should be treated with aminophylline (theophylline ethylenediamine). This may be injected slowly intravenously in doses of 0.25 to 0.5 gm or used in the same doses as a rectal suppository.

The most effective drugs for the relief of asthma are the adrenal corticosteroids (hydrocortisone and its derivatives) and corticotropin, which stimulates the endogenous secretion of corticosteroids. Because of the many side effects of prolonged use of these drugs in uncomplicated asthma, they are usually reserved for patients in whom all other types of treatment have failed. When plans for surgery make prompt control essential, their use for a few days or a week may be justified.

In acute status asthmaticus, corticosteroids, such as hydrocortisone, 100 mg intravenously, are the mainstay of treatment. When the condition is less urgent, the intramuscular injection of slowly absorbed corticosteroids, such as methylprednisolone acetate, 40 to 80 mg, may be useful. Asthma which has required repeated doses of epinephrine or aminophylline for several days is frequently controlled by oral administration of prednisone, starting with doses of 15 to 30 mg a day and tapering off in the course of a week.

Rapid control of bronchial asthma is aided by using intermittent positive pressure with inhalation of isoproterenol and a wetting agent such as 50% alcohol, usually for ten minutes four times a day. Such treatment aids in the elimination of mucoid sputum, but if it is particularly thick and tenacious, a saturated solution of potassium iodide, 10 drops three times daily should be given concurrently.

If fever, leucocytosis, or purulent sputum indicates the presence of infection, antibiotics are also indicated. The choice of antibiotics is best based on sensitivity tests of sputum cultures, but tetracycline, 250 mg, four times a day, may be started pending laboratory results.

By the use of such measures, most cases of asthma can be controlled within a few days to permit major surgery.

After operation, the patient is continued several days on antibiotics, and if the use of corticosteroids has proved necessary, they are gradually tapered off in the week after operation. During the postoperative period, the patient should be carefully watched for pulmonary complications.

Allergic Rhinitis

The preoperative care of patients with allergic rhinitis follows the same general principles outlined for asthma patients. Contact with extrinsic allergens is avoided by hospitalization in a clean, air-conditioned room for several days. The antihistamines are among the most valuable drugs in this disease. Usually one of the less sedative oral drugs such as chlorpheniramine maleate, 8 to 12 mg, three times a day, suffices. Pyribenzamine®, 50 mg, is more potent but more sedative. If even the more potent antihistamines such as Pyribenzamine and diphenhydramine are not effective, a corticosteroid such as prednisone, 10 to 15 mg daily, can be used concurrently.

Topical applications of adrenergic amines, such as phenylephrine hydrochloride and oxymetazoline hydrochloride, which often cause "rebound" congestion if used over a prolonged period, may be safely used for a few days before operation if necessary.

Allergic Dermatitis

Allergies of the skin only occasionally present a serious problem in surgery. In the treatment of atopic dermatitis, topical application of corticosteroids, such as hydrocortisone 1% or the newer synthetic derivatives in corresponding strengths are usually effective. Clearing the rash may be hastened by systemic medication with steroids. The intramuscular injection of methylprednisolone, 40 mg, usually produces a favorable response and may be followed by oral prednisone in doses of 10 to 20 mg a day for a week or so. If the patient has had no prior treatment with corticosteroids, this dosage does not present a problem at the time of operation.

Urticaria may usually be controlled with antihistamines, epinephrine and, when necessary, short courses of prednisone, 10 to 15 mg per day (Chap. 9).

Preoperative Studies of Patient Status

Preoperative studies of patients with allergic disease should include a careful search for evidences of infection that may affect the course of the allergy. In respiratory allergy, infections of the paranasal sinuses are particularly important, and in doubtful cases, x-ray films are needed to supplement the physical examination, since a bilateral involvement may be overlooked on transillumination. When the x-ray films suggest that fluid is present in the sinus, an exploratory puncture is desirable. If this yields purulent fluid, it is usually beneficial and should be repeated at intervals of two or three days until a clear fluid is returned. Such acute infections should also be treated with antibiotics and suitable local measures to aid drainage. Chronic hyperplastic sinusitis, common in allergic patients, is less susceptible to treatment. Its presence, or that of nasal polyps visible in the nares, has an important significance in the prognosis of the allergic disease. These hyperplastic changes associated with bronchial asthma generally suggest an allergic state difficult to control. It is considered suggestive, but not diagnostic, of violent sensitivity to aspirin.

If the patient has been treated with corticosteroid drugs for four or more weeks during the previous six months, the functional state of the adrenal cortex should be evaluated before operation. A number of suitable tests have been devised by Thorn[4] and others. Most of these measure the activity of the adrenal cortex when stimulated by injection of corticotropin. The total blood eosinophils normally decrease 50 percent in four hours after the intramuscular injection of 25 USP units of ACTH. More accurate tests involve chemical measures of the plasma and urinary levels of 17-hydroxycorticoids before and after the infusion of 25 units of ACTH over a period of eight hours.

If the results of such tests indicate that adrenal function is depressed, the possibility that the stress of operation may cause acute adrenal insufficiency must be recognized and suitable steps be taken to prevent its occurrence. In doubtful cases, consultation with an endocrinologist may be advisable. If such expert advice is not available, or the urgency of operation does not allow time for it, 100 to 200 mg of cortisone acetate should be injected intravenously before operation. During the operation, if signs of incipient shock, such as falling blood pressure appear, this dosage may be supplemented by the intravenous injection of soluble hydrocortisone. A considerable excess of corticosteroids during a few days of operative stress is generally less hazardous than an insufficiency.

Before operation, the anesthetist should be given all available information on the diagnosis and an evaluation of the severity of the patient's allergic disease. The choice of general anesthesia for the patient is discussed in Chapter 8.

The use of local anesthesia in the allergic patient has been the subject of conflicting opinions (Chaps. 8, 9).

However, the most serious and important hazards are immediate general reactions manifested by syncopy, convulsions and respiratory failure, occasionally fatal. Such reactions have followed both topical application and injection of cocaine, procaine and related synthetic drugs. Although the

abrupt onset and rapid progression of such reactions is superficially suggestive of anaphylactic shock and many writers have attributed them to allergy,
there is little real pathogenetic evidence that they result from specific
sensitizing action rather than from an unusual susceptibility to the normal
pharmacologic effects of the drug, overdosage, or rapid absorption into the
circulation. Such reactions are not limited to persons with other known
allergies; distinctive manifestations of allergy, such as urticaria, are usually
lacking.

Indeed, intravenous procaine has been suggested and used (rather in-
effectually) for the relief of bronchial asthma, serum sickness and other
allergic diseases.[5]

Despite the lack of conclusive evidence, however, many allergic patients
believe they are allergic to local anesthetics, and the calamitous reactions
reported have made physicians and dentists hesitate to disregard these
opinions. For an evaluation, allergists are frequently asked to perform skin
tests. Their general experience, despite occasional reports that such tests
are of diagnostic value,[6] is that neither scratch nor intracutaneous tests with
local anesthetics are of real value in predicting the possibility of reactions
or in the selection of the safest drug for a particular patient.

Goodman and Gilman[7] question the importance of allergy in such reactions and attribute them to excessive blood levels of the drugs used.
Dosage should be kept as low as possible, and accidental intravenous injection should be avoided. Injecting the anesthetic agent with epinephrine
helps to decrease absorption but may cause palpitations or other mild symptoms that increase the patient's anxiety. As preanesthetic medication, they
advise barbiturates such as phenobarbital and secobarbital, rather than
specifically antiallergic drugs.

Injection of Foreign Materials

The parenteral injection of any foreign material carries the hazard of
an anaphylactic, anaphylactoid, or general allergic reaction. These reactions
are always serious, difficult to treat and occasionally rapidly fatal. While not
limited to patients of known allergic background, they are much more
common in such patients, and special care is needed for those who have
previously shown evidence of an allergic tendency. In general, the severity
and danger of such reactions depend upon the nature and amount of the
foreign substance injected and the rapidity of its absorption into the circulation.

As a rule, foreign substances should be injected into the veins of allergic
patients only when this route is essential for the purpose intended. If intramuscular administration is suitable, injection into muscles of the thigh or

upper arm permits the effective application of a rubber tourniquet proximal to the site of injection to delay absorption. If a general reaction occurs, the tourniquet should be immediately applied and tightened to permit arterial flow of the blood to the extremity but prevent the return of venous blood. The initial treatment of an anaphylactic reaction should be epinephrine (Chap. 1).

The hazards of heterologous antisera are particularly great since the material is highly antigenic. In the past, most prophylactic and therapeutic antisera have been produced in horses. These often caused drastic reactions both in patients who had been sensitized by previous injections of horse serum and in persons with respiratory allergy following exposure to horses. Before using such materials, it is essential to evaluate the possibility of sensitization by both the history of previous use of serum and by intracutaneous tests with horse serum diluted 1 to 10. For a reliable test, 0.01 to 0.02 ml is injected as superficially as possible intracutaneously and the local reaction watched for fifteen minutes. A definite wheal 1 cm or more in diameter indicates that the patient is dangerously allergic to horse serum. If the injection is to be a prophylactic dose of tetanus antitoxin, material from the other sources should be used. If bovine serum is employed, the same tests for sensitivity should be made with it, but when human serum is used, skin tests are not advised.

Many of the less commonly used antitoxins (e.g. those for botulism and snake bites) are available only in horse serum. Patients unduly sensitive to horse serum, as a result of its previous use, can generally be desensitized over a period of several hours by a series of small doses injected at intervals of twenty to thirty minutes, starting with the amount injected in the skin test and doubling the dose each time. Such desensitization is rarely possible in the patient who is naturally highly sensitive and suffers respiratory symptoms from exposure to horses.

Contrast Media

Among the diagnostic procedures often needed in the preoperative period are the x-ray studies for which contrast media are injected intravenously—for pyelograms, angiograms and the like. The commonly used contrast media are complex organic compounds containing approximately 50% radiopaque iodine. For satisfactory x-ray films, it is often necessary to inject a large amount into the vein at a rapid rate. These injections occasionally cause general reactions. Although not limited to allergic patients, the hazard is definitely greater in patients with allergic diseases, particularly bronchial asthma. Whether these reactions result from specific allergy, from nonspecific release of histamine, or other mechanisms is not clear. Nor is their incidence significantly correlated with other evidences of allergy to iodides.

In the asthmatic patient, however, there is a serious risk of a severe asthmatic attack immediately after the injection. In the belief that this results from allergy to the injected agent, some authors have recommended skin or eye tests with the agent to be used, but, most workers have not found such tests reliable. The frequency of severe reactions and the lack of reliable tests to foresee them make intravenous pyelograms in asthmatic patients hazardous. They should be performed only when essential for planning treatment. In many cases, the safer retrograde pyelogram may be substituted.

When the intravenous method is considered essential in an asthmatic patient, the safest procedure is to inject a few drops of the contrast medium intravenously and watch several minutes for any evidence of reaction before injecting the total volume. The chance of a reaction may be decreased by giving 100 mg of diphenhydramine by mouth one hour before the procedure. The drug should not be injected with the contrast medium because it causes precipitation in some of these agents.

Penicillin

In performing surgical operations on allergic patients, it is well to avoid the prophylactic use of penicillin when this is not indicated by definite bacteriologic evidence of existing infection. Allergic reactions to penicillin are not limited to patients with other allergic diseases but the incidence in allergic patients, particularly those with bronchial asthma, is much higher than in the general population. A number of severe and even fatal reactions have followed the casual prophylactic injection of penicillin before or after operation in the absence of known infection or any specific indication.

Although skins tests with penicillin itself or penicilloyl-polylysine are of some value in diagnosing anaphylactic sensitivity to penicillin, no type of test is entirely reliable. The incidence of acute anaphylactic reactions is far greater with penicillin than with any of the other common antibiotics, so it is prudent to substitute another drug of the group for prophylactic use in patients with allergic disease.

Urgent Operations

When the need for surgical treatment is urgent, the timing of operation and the preoperative study advised in the case of elective operations are impossible. However, even in urgent cases, the existence of such serious allergic diseases as bronchial asthma and an estimate of their severity by the methods described in Chapter 4 must be made as rapidly as possible so that the surgeon and anesthetist are acquainted with the problems they face. It is particularly important to learn whether corticosteroid drugs have been used in the treatment of the allergic disease and, if so, the amount and duration of the dosage. When this information is incomplete or uncertain, the safest view is that use of corticosteroids over a period of more than two weeks within the previous four to six months suggests the possibility of suppression of the function of the adrenal cortex. This creates the risk of acute adrenal insufficiency with shock and collapse at the time of operation. In the absence of the usual contraindications, the corticosteroids may be used rather freely at the time of operation. The older drugs such as cortisone and hydrocortisone are preferable to the newer synthetic derivatives, many of which have been selected because of relatively slight effects upon salt, water and glucose metabolism which are essential in combating shock. The patient whose adrenal cortical function is uncertain should be protected by the injection of 100 to 200 mg of cortisone acetate intramuscularly before operation. This requires several hours to attain its full effect; therefore, if operation must be started at once, it should be supplemented by the intravenous injection of soluble hydrocortisone for prompt effect. After operation, intramuscular injection of cortisone acetate may be repeated daily until the vital signs are stable.

While this liberal use of cortisone is intended primarily to protect the patient against adrenal insufficiency during the period of operative stress, it usually has a concurrent favorable effect of temporarily suppressing the symptoms of allergic disease. When the dosage is tapered off after operation, allergic symptoms may flare up. These are then treated with suitable doses of the newer synthetic steroids. If the patient is then tolerating oral medication, prednisone is useful; if medication by mouth is still impossible, methylprednisolone intramuscularly may be used temporarily.

Operations on Tissues Affected by Allergic Disease

Operations on the tissues actually affected by allergic disease are most frequently necessary in procedures on the nose (Chap. 10). A majority of such operations are elective and when they are contemplated, it should be borne in mind that the resistance of the nasal mucosa to infection is greatly decreased by the allergic reaction. Postoperative healing may also be impaired. Finally, the anatomy of the nose may be so distorted by the edema of the allergic reaction that even the need for operation may be open to question.

In general, operations upon the noses of patients who are severely affected by allergic rhinitis should be delayed until the allergic component has been carefully studied and often until treatment of this aspect has been started. Patients with marked allergic rhinitis often also show mild deviations of the nasal septum. There may be difficulty in judging the relative importance of the allergic reaction and the septal deviation in causing obstruction to airflow. In such cases, careful allergic studies should be completed before surgical procedures are undertaken. If these studies show evidence of marked reaction to common allergens to which the patient is exposed, treatment with these may be carried out before a final decision as to the need for surgery.

REFERENCES

1. Sheldon, J.M., Lovell, R.C., and Mathews, K.P.: *A Manual of Clinical Allergy*, 2nd ed. Philadelphia, W.B. Saunders Co., 1967.
2. Sherman, W.B.: *Hypersensitivity: Mechanism and Management.* Philadelphia, W.B. Saunders Co., 1968.
3. Austen, K.F., and Becker, E.L.: *Biochemistry of the Acute Allergic Reaction.* Philadelphia, F.A. Davis Co., 1968.
4. Thorn, G.W., Forsham, P., and Emerson, K.: *The Diagnosis and Treatment of Adrenal Insufficiency*, 2nd ed. Springfield, Charles C Thomas, 1951.
5. Beckman, H.: *Pharmacology in Clinical Practice.* Philadelphia, W.B. Saunders Co., 1952, p. 15.
6. Aldrete, J.A., and Johnson, D.A.: Allergy to local anesthetics. *JAMA, 207:*356, 1969.
7. Goodman, L.S., and Gilman, A.: *Pharmacological Basis of Therapeutics*, 3rd ed. New York, Macmillan Co., 1965, p. 383.
8. Parker, C.W., Shapiro, J., Kern, M., and Eisen, H.N.: Hypersensitivity to pencillenic acid derivatives in human beings with penicillin allergy. *J Exp Med, 115:*821, 1962.

Chapter 3

Medical Management of the Allergic Child Before and After Surgery

Frederic Speer and James E. Shira

THE FIRST reaction of the surgeon and his associates to the fact that they are dealing with an allergic child is likely to be one of annoyance and vexation. They know that they must deal not only with the exigencies of surgery but the vagaries of allergy. Fortunately, we should hasten to say, allergic individuals are much more resistant to such stresses as surgery than they seem. When we take note of their sensitivities and consider the limitations imposed by such problems as asthma, they do as well as other surgical patients.

The only way to avoid trouble with allergic children is to assume that every child is potentially allergic. In other words, it is the "unknown allergic" who is likely to cause trouble. In a review of incidence of allergy, Kantor[1] found that minor allergy may occur in from 20 to 50 percent of the population. Such minor allergy may at any time become an important factor in a patient under stress. We will therefore begin our chapter with a method of screening children so as to find allergic traits and possible allergens.

IDENTIFYING THE ALLERGIC CHILD

Data from the Referring Physician

The child's own physician will want to inform the surgical team about any illnesses or weakness that may be of significance during surgery and convalescence. Inquiry should always be made. Since the referring physician may not consider minor manifestations (like contact dermatitis from adhesive tape) to be important, it is well to ask about *all* allergic traits, minor as well as major. At the same time the referring physician may outline precautions that will need to be considered, such as special treatment for asthma, avoidance of certain drugs and foods, and care of allergic dermatoses.

Data from the Patient and His Family

Whenever any child is admitted to the hospital, questioning members of the family as to possible allergy should be standard practice. In Appendix 1 to this chapter is shown a form that may be used for this purpose. To it may be added questions as to other phases of this history such as known diseases, toilet habits and emotional needs.

DEALING WITH ALLERGENS

Inhalants

The common and important inhalants are pollens, molds, animal danders, feathers, insect sprays and house dust. Minor offenders like cottonseed, kapok, flaxseed, orrisroot and jute seldom are important in hospital patients. If the child's room is not air-conditioned, he will at certain seasons be exposed to air-borne molds and pollens. However, some pollen and mold spores are certain to be present, even in carefully air-conditioned rooms. Therefore, children suffering from poorly controlled pollen allergy should not have elective surgery during the pollen season; this is especially true of allergy to ragweed and other weeds. The same restrictions may apply to children with mold allergy, especially in agricultural areas.

The hospitalized allergic child should have no trouble from inhalants if three conditions are met: (a) an air-conditioned room, (b) nonallergenic bedding, and (c) exclusion of fumes or odors that may cause respiratory symptoms. Fortunately, hospital mattresses and pillows are usually covered with heavy rubber and are therefore dust-free. If, however, there is a feather or down pillow, it must be replaced with one made of synthetic materials such as Orlon®, a foam rubber pillow, or a pillow covered with plastic or rubber. A cotton mattress may be covered with impervious material or replaced by one of foam rubber.

Although true inhalant allergens like pollens, molds, animal epithelials and house dust are the chief source of respiratory allergy, fumes and odors that the child may encounter in the hospital are a more immediate threat.[2] Tobacco smoke is especially important, and the child should not be in a ward with smokers or with patients whose visitors are allowed to smoke.[3] Other common offenders are perfumed toilet articles, paint, tincture of green soap, room deodorants, smog and other air pollutants (even in air-conditioned rooms),[4] Christmas trees and decorations, chlorine bleaches, Lysol®, floor wax, furniture polish, mothballs, insect sprays, rubbing alcohol, hair sprays and flowers. Many an allergic patient has suffered severely from the flowers and smoke introduced into his hospital room by unthinking though well-meaning visitors.

Avoiding Food Allergens

When a food allergen is known to cause the child trouble, it is the responsibility of the hospital not to serve him that food. In this case, the child's mother can usually furnish the dietician with lists of foods that must be avoided and suggest suitable and palatable substitutes. Where such information is not available, the lists in Appendix 2 of this chapter can be used. The food allergens are listed in descending sequential importance as causes of allergy in children. Since the child is not likely to be in the hospital for more than a few days, elaborate recipes for preparing meals without such foods as egg and wheat usually need not be used. If the hospitalization is to be prolonged, the dietician can consult a textbook for guidance.*

In most cases, reactions to foods are not severe, but they are always a threat, especially to young children. The potential seriousness of such reactions is reemphasized in a recent paper by Gryboski.[5] She reports seven cases of cardiovascular collapse caused by allergy to milk.

DEALING WITH REACTIONS TO THERAPY

Antibiotics

Unquestionably the chief problem in the management of the hospitalized allergic child is the possibility of a reaction to an antibiotic. Fortunately the characteristics (if not always the mechanisms) of these reactions are well known. We may therefore take proper precautions to avoid at least the more serious reactions.

Penicillin

Among the antibiotics, the most common and most serious offender is penicillin. It is impossible for the surgeon (or anyone else) to foresee and avoid all reactions to penicillin, but he must at all costs be alert to the serious possibility of an anaphylactic reaction. Such a disaster is unlikely to occur if history-taking has been thorough. Even in the absence of a history of a reaction (usually urticaria) the drug should always be used with care. It may cause reactions by injection, oral administration,[6] or even by inhalation.[7] Epinephrine should always be at hand when the drug is given parenterally.

Allergists have long been frustrated by the fact that skin testing for penicillin is neither safe nor dependable.[8-10] Even a scratch test may cause anaphylactic reactions, as illustrated by a recent fatal case.[11] In this instance the patient gave a history of probable allergy to penicillin. Although under observation and promptly treated with epinephrine, the patient died five minutes after having been given a scratch test containing less than one unit of the drug. Our best hope for the future appears to be the development of

---

*Any of the following books would be good:

(1) Conrad, M.L.: *Allergy Cooking*. New York, Thomas Y. Crowell, 1960; (b) Rinkel, H.J. *et al.*: *Food Allergy*. Springfield, Charles C Thomas, 1950; (c) Rowe, A.H.: *Elimination Diets and the Patient's Allergies*. Philadelphia, Lea & Febiger, 1944; or (d) Speer, F.: *The Allergic Child*. New York, Hoeber, 1961.

a dependable *in vitro* test. We must at present depend entirely on the history.

Other Antimicrobials

In the following outline are listed antibiotics in common use and their common side reactions. The outline is based on the recommendations of *New Drugs*[12] and the current literature.

Antibiotic	Side Effects
Chloramphenicol	Aplastic anemia. Drug should be used only when absolutely necessary for organisms known to be susceptible to it.
Colistins	Occasional neurotoxicity; nephrotubular degeneration may occur in infants. Restrict use to infections refractory to usual agents.
Erythromycins	Seldom cause allergic reactions. Since the estolate may cause jaundice, it is contraindicated in case of impaired liver function.
Tetracyclines	Dental damage in young children; superinfection with G.I. and vaginal complications. Demethylchlortetracycline (Declomycin®) common cause of photosensitivity.
Nitrofurantoin	Pulmonary infiltration with fever, much like pneumonia[13]; nausea, vomiting, various dermatoses.
Sulfonamides	Crystalluria possible, especially in oliguria; rashes, drug fever, photosensitivity.
Streptomycin and kanamycin	Injury to eighth nerve.
Lincomycin	Reactions to this new drug minor, to date.
Nalidixic acid	Occasional nausea, urticaria, other rashes.
Novobiocins	Sensitization common, manifestations diverse. To be avoided in prematures and newborns.
Cephalothin	Cross-reacts with penicillin, at least to some extent.[14,15]
Vancomycin	Reactions unusual unless renal impairment present.
Triacetyloleandomycin (Tao®)	Prolonged use may cause decreased hepatic function.
Nystatin	Reactions mild and rare.
Griseofulvin	Occasional skin allergies and constitutional symptoms.
Amphotericin B	May cause permanent kidney damage. Restrict to patients with potentially fatal mycotic infections.

Drugs That Affect Mood

This category includes *analeptics* (methylphenidate, ethamivin, amphetamines), *antidepressants* (amitriptyline, phenelzine, MAO inhibitors), *tranquilizers, sedatives and hypnotics*. Allergic children are remarkably intolerant of most of these drugs, and the patient usually feels worse after taking the drug than he did before. If sedation is required, most children tolerate chloral hydrate or diphenhydramine hydrochloride. In more severe cases, such narcotics as morphine and meperidine hydrochloride (Demerol®) seem to be as well tolerated by allergic children as by others. These drugs are, however, strictly contraindicated during an asthma attack.

Analgesics

Except for allergy to aspirin, this class of drugs is well tolerated by most allergic children. Since aspirin allergy may be severe, it should never be given to an allergic patient until inquiry has been made into previous reactions. If they have occurred, a prominent notation must then be entered on the chart that no aspirin or aspirin-containing preparations are to be given. Acetaminophen (Tylenol, Tempra®) is useful in children allergic to aspirin. Both codeine and dextropropoxyphene (Darvon®) are usually well tolerated by allergic patients, but it must not be forgotten that many analgesic mixtures contain aspirin.

Bronchodilators

Although epinephrine comes to mind first as therapy for asthma, in some children alarming shakiness, pallor, or even faintness may develop after an injection. These children almost always can take ethylnorepinephrine (Bronkephrine®). Children who do not tolerate ephedrine usually can take racephedrine-containing products like Ephoxamine® and Amodrine®. Unless given in excessive dosage, aminophylline is a safe and effective drug in asthma.

Immunizing Agents

All drugs under this heading can cause serious reactions, especially in allergic children. Along with penicillin and aspirin, they should always be viewed with suspicion. The following agents are those most likely to be given to hospitalized children.

Tetanus Antiserum

The dangers of horse serum are well known. The physician is wise to follow the ultracareful scheme of Feinberg.[16] Steps in testing are as follows: (a) scratch test with a 1:1000 dilution, intradermal testing at 15-minute intervals with (b) 1:1000, (c) 1:10, and (d) undiluted serum. Testing is stopped at any point when there is a wheal and flare reaction, and the serum is not used. Undoubtedly most physicians now prefer using human tetanus immune globulin, since the possibility of a reaction is remote. Needless to say, most children have had recent immunization by tetanus toxoid and are given a booster dose rather than the serum. The same precautions that are followed here also apply to polyvalent gas gangrene horse serum.

Tetanus Toxoid

Although reactions to toxoid are less severe than those to the serum, they may occur. To reduce their incidence, Edsall *et al.*[17] recommend that emergency recall doses be given no oftener than annually and that boosters be given no oftener than every ten years. The reactions may be allergic but are more commonly of the Arthus type, which may include inflammation, edema, hemorrhage and necrosis. Children who have had large local reactions should not be given another dose but should have the human tetanus immune globulin instead.

Diphtheria Antiserum

The same precautions apply to the use of diphtheria antiserum as apply to tetanus antiserum. So-called rapid desensitization may be required when the use of serum is definitely indicated.

Influenza Vaccines

These vaccines may be indicated in the face of an epidemic. Children extremely allergic to egg may not tolerate them nor other vaccines grown on chick embryo. Other vaccines that may be grown on chick embryo are those against yellow fever, mumps, measles, Rocky Mountain spotted fever and epidermic typus. A preliminary skin test is indicated when doubt exists.

Rabies Vaccine

The danger of neurologic complications developing following rabies prophylaxis has been greatly decreased by the introduction of duck embryo vaccine.[18] Since some local and systemic reactions to this vaccine may occur, it should be used with caution, especially in patients known to be allergic to egg.

Antivenin

For some reason, reactions to horse serum against the venoms of pit vipers are especially common and severe. Parrish[19] recommends that the physician first make certain that the bite is actually that of one of these snakes. The same precautions as in other types of horse serum preparations must then be followed.

Miscellaneous Agents

Enzymes

According to *New Drugs*,[12] anaphylactic reactions are a strong possibility following administration of chymotrypsin, streptokinase, streptodornase and penicillinase. We may expect such reactions to be especially common in allergic children.

Contrast Media

Iodinated organic compounds are widely used as radiopaque media in diagnostic studies in children with urologic problems. These agents are histamine liberators and are thought to be especially dangerous in allergic individuals. The need of careful inquiry to detect any history of preceding reactions is imperative.

Blood Substitutes

Plasma infusions seem to be no more hazardous in allergic children than in other patients. The dextrans, however, seem to be rather common causes of anaphylactic reactions.[20,21] It would therefore seem wise to make every effort to avoid their use in allergic patients.

Gamma Globulin

There should be few indications for gamma globulin in children hospitalized for surgery. Its use as "shotgun" treatment for prevention and treatment of infection is not justified. Allergic reactions are a possibility, especially in children with hypogammaglobulinemia.[22,23]

MEDICAL CARE OF THE CHILD WITH CONTROLLED ASTHMA

The one allergic disease both physicians and surgeons must treat with the utmost respect is asthma. The youngster with asthma under good medical control tolerates the rigors of anesthesia, surgery and postoperative convalescence as would any other child. But if his medical support is relaxed, he may well have an attack of asthma or even go into respiratory arrest during induction of anesthesia, during the operation, or in the recovery room. At the least, his convalescence may be prolonged because of continuing respiratory distress from uncontrolled allergic bronchial disease. The asthmatic child's needs during hospitalization must therefore be anticipated and the necessary prophylactic steps be taken prior to admission.

Management of asthma as well as any allergic disease in children revolves around three basic therapeutic measures: (a) removal of known and potentially causative allergens through dietary eliminations and environmental control measures; (b) enhancement of the patient's tolerance of known inhalant allergens through hyposensitization injections; and (c) prevention or relief of symptoms with appropriate medications.

Measures to control environment and diet have already been presented. However, several other practical considerations also arise.

Scheduling Hospitalization for Surgery

Elective surgery, by definition, is scheduled when conditions are as close to optimum as possible. Thus, for the child who is experiencing a flare of his allergic disease, deferment of operation is recommended until his symptoms can be resolved. Likewise, surgery on the child with poorly controlled pollinosis should be postponed until his pollen season is well past.

Continuation of Medications for Asthma Control

Most children with asthma do not require maintenance medication. Between their acute episodes of dyspnea and wheezing they are symptom-free, enjoy a full spectrum of activity and have normal pulmonary function. They may, of course, have collateral allergic disease such as rhinitis, hives, or eczema that demands long-term treatment. Some asthmatics, on the other hand, definitely need daily medications to control allergies of their lower respiratory tracts; such medication should be maintained during the entire hospitalization to insure that their asthma remains quiescent. Table 3-1 lists those drugs commonly used in the outpatient management of childhood asthma. The doses recommended must be considered only as guidelines,

TABLE 3-1
DRUGS FOR MAINTENANCE TREATMENT OF ASTHMA

Drugs	How Supplied		Dose	Maximal Dose
Bronchodilators				
A. Sympathomimetics				
1. Ephedrine	Capsule	25 mg		
	Syrup (NF)	20 mg/5 cc	4-5 mg/kg/24 hr	25 mg q 4 h
	Syrup (USP)	10 mg/5 cc	in 4 to 6 doses	
2. Pseudoephedrine	Tablets	30 mg and		
		60 mg	5 mg/kg/24 hr	60 mg q 6 h
	Syrup	30 mg/5 cc	in 4 to 6 doses	
3. Propadrine	Capsules	25 mg and		
		50 mg	5 mg/kg/24 hr	50 mg q 6 h
	Elixir	25 mg/5 cc	in 4 to 6 doses	
B. Xanthines				
1. Theophylline	Tablets	80 mg and		
		100 mg	12-15 mg/kg/24 hr	240 mg q 6 h
	Elixir	27 mg/5 cc	in 4 to 6 doses	
	Rectal units			
	500 and 250 mg		12 mg/kg/24 hr	250 mg q 8 h
			in 3 doses	
2. Oxtriphylline	Tablets	100 mg and		
		200 mg	15 mg/kg/24 hr	200 mg q 6 h
	Elixir	100 mg/5 cc	in 4 to 6 doses	
3. Aminophylline	Suppository			
	125, 250 and 500 mg		20 mg/kg/24 hr	250 mg q 8 h
			in 3 doses	
Expectorants				
A. Iodides				
1. Potassium iodide	Saturated solution		1 drop/year of	10 drops q.i.d.
			age, q.i.d.	
2. Organidin	Tablets	15 mg	½-1 tablet q.i.d.	1 tablet q.i.d.
(iodinated)	Drops	25 mg/cc	5-10 drops	10 drops q.i.d.
glycerol)	Elixir	30 mg/5 cc	¼-½ teaspoon	½ teaspoon q.i.d.
3. Hydriodic acid	Syrup (NF)	70 mg/5 cc	5-10 cc q.i.d. or	10 cc q.i.d.
			q 6 h	
B. Glyceryl guaiacolate	Syrup	100 mg/5 cc	2.5-10 cc q.i.d.	10 cc q.i.d.

since each patient varies in his response to, and tolerance of, the medications. Here again, the patient's referring physician can provide the surgeon with invaluable information concerning the best program of drugs for his patient.

Bronchodilators

The sympathomimetic and xanthine drugs, alone or in combination, are the mainstays of symptomatic treatment of asthma. Of the former group, ephedrine is particularly effective and has the advantages of being inexpensive and almost universally available. Because it may produce significant central nervous system stimulation in some children, ephedrine is often combined with a sedative such as phenobarbital. Agitation may also be reduced by substituting ephedrine with pseudoephedrine, racephedrine, Propadrine® hydrochloride or Orthoxine®. These agents, however, offer no other advantages as bronchodilators.

Theophylline, aminophylline and oxtriphylline are the xanthines most frequently prescribed for asthma. They may be administered orally, rectally (Table 3-1), or intravenously. Intravenous doses will be discussed later in this chapter with the treatment of the asthma attack. Like the sympathomimetics, the xanthines are effective bronchial relaxants but may produce central nervous system stimulation, cardiac arrhythmias, nausea, vomiting and hematemsis.

Many of the preparations now available combine expectorants and sedatives with ephedrine and theophylline, or with theophylline alone. These are primarily convenience preparations and offer no other advantage over using the ingredient drugs separately. Table 3-2 lists a few of the commonly used mixtures and their component drugs.

Expectorants

Children with asthma must be encouraged to maintain a large daily fluid intake to insure adequate liquification of secretions in the airway. Water remains the best expectorant available and is an essential prerequisite for any of the preparations mentioned below. In a small but definite segment of the asthmatic population, iodides are effective[24] and can be prescribed as saturated solution of potassium iodide, iodonated glycerol, or syrup of hydriodic acid. Glyceryl guaiacolate preparations may be of value when iodides are contraindicated or ineffective.[25] Acetylcysteine (N-acetyl-L-cysteine) is a potent mucolytic agent; however, it may cause prolonged bronchospasm and therefore is generally not prescribed for asthmatic patients.[26]

TABLE 3-2
SYMPATHOMIMETIC AND XANTHINE MIXTURES FOR ASTHMA THERAPY*
COMPONENTS IN MILLIGRAMS PER 5/CC (1 TSP)

Name	Ephedrine	Theophylline	Oxtriphylline	Phenobarbital	Iodide	Glyceryl Guaiacolate	Other
Asbron†		50 mg				33 mg	Phenylpropanolamine 8 mg
Brondecon‡			100 mg			50 mg	
Bronkolixir	12 mg	15 mg		4 mg		50 mg	Chlorpheniramine 1 mg
Elixophylline-KI		27 mg			43 mg		
Luasmin	15 mg	100 mg		15 mg			
Marax§ (½ strength tablet)	6.25 mg	32.5 mg					Hydroxyzine 2.5 mg
Quadrinal‡	12 mg	65 mg		12 mg	150 mg	30 mg	
Quibron†		50 mg					
Tedral‡	12 mg	65 mg		4 mg			
Verequad‡	12 mg	65 mg		4 mg		50 mg	

*Liquid preparations.

†Tablet or capsule equals 15 cc.

‡Tablet or capsule equals 10 cc.

§Tablet equals 20 cc.

Antibiotics

Since asthma and respiratory tract infections do not invariably go hand-in-hand, blanket antibiotic coverage of all asthmatic children undergoing surgery is not indicated. As is true of any other situation, respiratory tract infection in the allergic child contraindicates elective surgery. Further, since the asthmatic patient may well have a prolonged convalescence following respiratory tract infection, postponement of surgery may be longer than usual.

For a few children with chronic, unrelenting asthma significantly complicated by respiratory tract infection, prolonged courses of antibiotics may become a therapeutic necessity. When such a patient is admitted for operative care, his drug therapy must obviously be reassessed by the surgeon and his medical consultant. In most cases the antibiotic will be continued throughout the patient's entire hospitalization. In this situation, of course, one must be concerned about the emergence of resistant or opportunistic organisms such as *Candida, Staphylococcus, Pseudomonas,* or *Aerobacter.* The choice of antibiotics depends upon the patient's age, his clinical picture, the operation to be performed, drug sensitivities and the experience of the prescribing physician. In general, in the pediatric age group, penicillin, erythromycin, ampicillin or combinations of penicillin with a sulfonamide are commonly used. Erythromycin, sodium cephalothin, or lincomycin are indicated if penicillin sensitivity exists.

Corticosteroids

Although there is continuing debate concerning the indications for the use of steroids in the overall management of asthma,[27] there is general agreement that certain asthmatic children undergoing surgery demand corticosteroid coverage.[28-31] They are as follows:

1. The actively wheezing patient who must undergo emergency surgery and whose response to conventional medications (epinephrine and aminophylline) is less than complete.

2. The chronically ill patient who requires maintenance steroids for asthma control, including children on intermittent, every-other-day, or daily dosage schedules.

3. The patient who has received repeated short courses (six days or less) of corticosteroids for control of severe, acute asthmatic attacks during the year prior to admission.

4. The patient who has suspected or chemically documented suppression of the hypothalamic-pituitary-adrenal axis due to prior steroid therapy for any condition. In a child, significant suppression of the pituitary and adrenal gland may persist for up to twelve months following a three to four

week course of prednisone in a dosage of 5 to 10 mg or more daily. If prednisone or its equivalent is continued for three months on a daily basis, suppression could last for two years or more.[31-33]

Table 3-3 provides suggested dosage schedules for steroid administration at the time of surgery. Corticotropin (ACTH) is not included in the table and, generally, should not be prescribed. The onset of its action may be delayed and its effect unpredictable.[34,35] Since it has a foreign protein source, it may lead to serious sensitization reactions.[36]

Administration of Allergy Extracts

For most allergic children, the schedule of allergen hyposensitization injections is not so critical as to demand that the shots be given during a short hospitalization for surgery. If, however, the child is scheduled for a prolonged surgical admission, administration of these injections during the hospitalization may be necessary to avoid the emergence of symptoms. This is particularly true if the surgery is to be performed during the patient's pollen season or if the admission is during that period when the patient is undergoing a critical increase in his allergy extract dosage. Arrangements for the hyposensitization injections can be made with the child's allergist.

ACUTE ASTHMA

Since treatment of asthma is more effective when instituted early, particularly prior to the development of complications such as dehydration and acidosis, the hospital staff caring for an asthmatic child must be aware of the initial features of the disease so that prompt and proper therapy may be provided. The following are a few of the early, sometimes subtle, signs and symptoms of an asthmatic attack that may go unnoticed on a busy surgical ward unless the child is observed closely:

Frequent clearing of the throat
Sitting in a hunched position with shoulders elevated and pulled forward
Increase in the anteroposterior chest diameter
Shallow tachypnea
Dry or "croupy" cough
Rattling respirations
Nasal congestion and rhinorrhea
Frequent sneezing
Vomiting, particularly after coughing paroxysm

Once the attack of asthma has begun, the child's distress is obvious. He has gross wheezing and dyspnea, cough, tachypnea and, at times, grunting or sighing respirations. The wheezing infant may well appear more comfortable while flat on his back.[37] The older child, however, experiences

TABLE 3-3

CORTICOSTEROID COVERAGE FOR THE ASTHMATIC CHILD UNDERGOING SURGERY

Drug	Loading (Day Prior to Surgery)	Doses Operative (Day of Surgery)	Convalescent (Days after Surgery)
Hydrocortisone	40 to 80 mg P.O. in 4 doses or I.M. in 1 dose in AM	100 mg, IV, during surgery and 50 mg in Recovery	1st day — 100 mg 2nd day — 50 mg* 3rd day — 20 mg
Prednisone	10 to 30 mg P.O. in 3 doses or I.M. in 1 dose in AM	25 mg, IV, during surgery and 10 mg in Recovery	1st day — 25 mg 2nd day — 15 mg* 3rd day — 5 mg
Dexamethasone	1.5 to 3 mg P.O. in 3 doses or I.M. in 1 dose in AM	3 mg, IV, during surgery; 1 mg in Recovery	1st day — 3 mg 2nd day — 1.5 mg 3rd day — .75 mg

*Second and third day doses may be omitted with uneventful convalescence.

further distress when lying down and tends to seek the upright position for more comfortable breathing. If the attack progresses in severity, marked restlessness or anxiety develops. Cyanosis is uncommon until the attack is grave; pallor is frequent. The patient may talk in spurts during his prolonged periods of exhalation. Wheezing is predominantly, but not exclusively, expiratory. The child's chest is hyperexpanded and the percussion note is generally hyperresonant. Cardiac dullness is usually not percussible. Musical rales and rhonchi are universally present. The rales may be sticky and moist in character, particularly if infection is present. With pneumonia, an area of dullness may be discernible by percussion, but evidence is frequently absent on physical examination, and a chest x-ray film may be obtained.

The differential diagnosis of wheezing in the child can be rather long and complicated, as indicated in Table 3-4. In most situations, the diagnostic possibilities can rapidly be narrowed down to a few practical considerations. Aside from allergic asthma, wheezing occurring in the youngster hospitalized on a surgical ward should certainly provoke consideration of the following processes: bronchitis, pneumonia, foreign body aspiration, or postoperative respiratory insults such as atelectasis or pneumothorax. It is dangerous, of course, to assume that all that wheezes is asthma. At the same time, however, one must realize that most wheezing truly is due to allergic bronchial disease.

Treatment of the Child with an Asthmatic Attack

Treatment of the asthmatic attack is directed towards reversing the following underlying pathophysiologic processes: (a) bronchial muscle spasm; (b) edema of the bronchial mucosa; and (c) collection of mucous secretions in the bronchial lumina. In addition, therapy must be designed to correct complications of asthma, such as pneumonia or alterations of the body fluid and electrolyte equilibrium. In general, management of the uncomplicated attacks is simple and relief prompt. If the child does not respond to the usual measures as outlined below, the surgeon may wisely choose to consult with a pediatrician or allergist.

Drugs Effective Primarily in the Relief of Bronchospasm

Epinephrine, aqueous, 1:1000, is a potent bronchodilator and the drug of choice for the treatment of acute asthma. It is administered subcutaneously in a dosage of 0.1 to 0.3 ml every fifteen to twenty minutes for three or four doses. Most children require no more than 0.1 ml for prompt response. Although relief is usually gained from epinephrine within five to ten minutes after administration, symptoms may well return after a 20- to 30-

TABLE 3-4

DIFFERENTIAL DIAGNOSIS OF WHEEZING IN A CHILD

I. Congenital anomalies (of importance primarily in the wheezing infant)
  A. Respiratory Tract
    1. Choanal stenosis
    2. Redundant epiglottis
    3. Laryngomalacia, tracheomalacia, or bronchomalacia
      a. Laryngeal or tracheal webs
      b. Laryngeal, tracheal, bronchogenic, or lung cysts
      c. Tracheoesophageal fistula, H-type
  B. Vascular
    1. Double aortic arch
    2. Aberrant azygous or subclavian veins
    3. Anomalous pulmonary venous return

II. Inflammatory diseases
  A. Granulomatous
    1. Tuberculous pneumonia, bronchitis, or lymphadenitis
    2. Histoplasma pneumonia or mediastinitis
    3. Coccidiodomycosis pneumonia
  B. Pyogenic or viral
    Pneumonia, bronchiolitis, bronchitis, bronchiectasis,
    laryngotracheobronchitis, epiglottitis, pharyngitis
  C. Metazoan
    Ascariasis and toxocariasis — Loeffler's syndrome
  D. Protozoan
    Toxoplasmosis pneumonia or lymphadenitis
  E. Chemical (fumes, odors, gases and dust)
    Pneumonia, bronchitis, laryngotracheitis, rhinitis
  F. Collagen disorders
    Systemic lupus erythematosis

III. Neoplasms
  A. Carcinoma or sarcoma at any point along the respiratory tract
  B. Neuroblastoma, lymphoma and hamartoma — mediastinal or cervical
  C. Leukemic infiltration of the respiratory tract or associated lymphoid
    structures

IV. Diseases of obscure etiology or pathophysiology
  A. Cystic fibrosis
  B. Pulmonary hemosiderosis — Heiner's syndrome
  C. Diffuse pulmonary intersititial fibrosis—Hamman-Rich syndrome
  D. Immune deficiency states

V. Foreign body aspiration — esophagus and at any level of the respiratory tree

VI. Dyspnea associated with congestive heart failure

VII. Neurogenic dyspnea
  A. Postencephalitis hyperventilation
  B. Recurrent laryngeal nerve paralysis
  C. Bronchospasm of tetany

minute period. In most cases, epinephrine is the only drug that is necessary for the treatment of the acute asthmatic attack. It can then be followed with oral medications as outlined in previous sections of this chapter.

Aqueous epinephrine suspension (Sus-Phrine®) , 1:200, may be utilized to achieve a sustained sympathomimetic effect lasting up to six to eight hours. Dosage in children is 0.05 to 0.15 ml subcutaneously every six hours. In general, this medication should not be given to the patient until he has shown responsiveness to aqueous epinephrine.

Ethylnorepinephrine (Bronkephrine) may be helpful for those few children who are unable to tolerate epinephrine even in very small doses. It can be administered in the dose of 0.1 to 0.3 ml every twenty minutes for three or four doses.

Overdosage with sympathomimetic medications may lead to distressing restlessness, anxiety, headache, tachycardia, hypertension and cardiac arrhythmias.

Aminophylline is an ideal companion to epinephrine and is indicated when sympathomimetics fail. For the treatment of acute asthma, intravenous administration is preferred. The drug is also well absorbed through the rectal mucosa and may be given by this route if the patient's condition is not pressing. Intravenously, the dose for aminophylline is 4 mg/kg to be given over a 20-minute period (no faster!) and no more frequently than every eight hours. When the drug is being administered intravenously, a physician should be present to monitor the cardiac rate and rhythm. The total daily intravenous dose should not exceed 12 mg/kg of body weight. The dose for rectal administration of aminophylline varies according to the preparation used. With suppositories, 5 to 8 mg/kg every eight hours is safe. If rectal solutions are used, however, the maximal dose should be 5 mg/kg every eight hours.

As it is true with epinephrine, the physician must practice great caution with xanthines. If the child has been taking them or products containing them, further treatment with aminophylline may lead to acute intoxication. The toxic effects are striking and include vomiting, hemorrhagic gastritis, central nervous system irritability, convulsions, coma, cardiac arrhythmias and cardiac arrest.[38] Aminophylline has been referred to as the "asthmatic's poison." This is no more true of this drug than any other if the preparation is properly administered in the correct dose.

Corticosteroids, although beset with many potential hazards,[39] play a major role in the treatment of recalcitrant asthma. When administered for brief periods of time (5 to 7 days), their benefits far outweigh any possible side effects. Although debate continues over the relative merits of corticotropin (ACTH) as opposed to the adrenal steroid preparations, the latter are generally preferred.[40,41] The two preparations most commonly used today are hydrocortisone and dexamethasone. The dosage of hydrocortisone is 100 to 200 mg intravenously given immediately and then 25 to 50 mg intramuscularly every six hours for as long as necessary. Higher doses may be used if the patient's condition is critical, but they are seldom needed.

Dexamethasone is given immediately in a dose of 3 to 5 mg intravenously followed by 1 to 3 mg intramuscularly every six to eight hours. If steroids are required for a short period of time (3 to 5 days), there is no need to taper the doses. If the patient requires more than two weeks of therapy, however, it is recommended that the doses be tapered off over the course of another week.

It is most important to note that four to eight hours may pass before the therapeutic effects of the corticosteroids may become apparent. During this time it is very essential that the patient receive other supportive measures such as intravenous fluids, inhalation therapy and rest.

Control of Infection

The usual signs of infection such as fever and leukocytosis are not reliable indicators of infection in asthma. Changes in the character of the patient's sputum from clear and watery to thick and yellow-green may be a clue. Infiltrative processes on x-ray examination may represent pneumonia or patches of atelectasis. Choice of the therapeutic agent is governed by the clinical picture. Obviously, before any antibiotics are started, cultures of the airway secretions and blood should be obtained. When the physician must resort to empiric antibiotic therapy, coverage should be made for *Streptococcus, Pneumonococcus* and *Haemophilus influenzae.* These organisms are best treated with either ampicillin or the tetracyclines. If a specific organism can be grown, sensitivities will then dictate the antimicrobial agent to be used.

Use of Sedatives

Extreme caution must be exercised when considering sedation for the asthmatic patient.[48] Irritability, anxiety and hyperactivity are all symptoms of hypoxia as well as emotional distress. Oxygen may take care of all these symptoms. Morphine and meperidine are strictly contraindicated except for those patients placed on controlled ventilation therapy. Barbiturates, even when prescribed in the usual dosages, may significantly depress the respiratory center and therefore should not be given. Tranquilizers such as chlorpromazine and hydroxyzine possess atropine-like activity and tend to thicken airway secretions. Only rarely is a drug such as chloral hydrate useful in the management of acute asthma. The dose is 15 mg/kg orally every six hours.

Fluid and Electrolyte Administration

The child with asthma may become significantly dehydrated, even while in the hospital, because of poor fluid intake coupled with excessive fluid losses from hyperventilation, perspiration and vomiting. If the symptoms are severe enough, it is very likely that the child's anorexia will not permit him to maintain an adequate fluid intake by mouth. Intravenous fluids then become imperative. The same principles that apply to fluid and electrolyte administration in other diseases are applicable in the treatment of asthma: replacement of maintenance requirements; replacement of prior and concurrent losses of water and salt; and repair of acid-base imbalance. Usually, water is the asthmatic's greatest need. His dehydration seldom varies from isotonicity. If no signs of dehydration are clinically apparent, it is safe to assume that the patient is 3 to 5 percent deficient in body water. This assumption will insure adequate fluid administration for liquification, mobilization of airway mucus and replacement of required fluid. Electrolyte administration is obviously dependent on the amount of fluid given, the urinary output and serial determinations of serum electrolyte concentrations. Since most children with asthma have normal renal function, fluid and electrolyte administration can be simplified with the use of polyionic-hypotonic solutions.

Mist Therapy

Most asthmatic children improve significantly when placed in a moist atmosphere. The inhaled water helps to thin airway secretions, promote mobilization of thickened mucus, and prevent excessive loss of water from hyperventilation. A number of efficient nebulization units are now available for humidification of tent structures.[42] Recently more and more emphasis has been placed on ultrasonic nebulizers, which reportedly have the advantage of producing a uniform, small droplet mist.[43] The condition of a few asthmatic children (perhaps 10% to 15%) consistently worsens when placed in high humidity. Obviously, for those children, mist therapy may not be desirable.

Intermittent aerosol therapy utilizing a pump-driven nebulizer, ultrasonic nebulizer, or positive pressure ventilation can be a helpful adjunct to the already mentioned therapeutic measures. Isoproterenol 1:300, 5 drops in 2 ml of normal saline, may be used as an aerosol solution. This drug is a potent bronchodilator and can be administered in this dosage safely every four to six hours for two days.

Caution

The excessive inhalation of isoproterenol may lead to significant rebound bronchospasm and refractiveness to other forms of therapy.[44-46] Sudden death from cardiac arrest has been reported when epinephrine and isoproterenol have been administered together, particularly to a patient who is hypoxic.[47] If the child is in the habit of using a hand nebulizer, he should not be allowed to use it at his own discretion. If he seems unable to do without it, it may be ordered by the doctor and administered under the direction of a nurse. Each nebulizer has package inserts giving dosage and intervals between uses. In view of recent reports on harmful effects of this type of treatment, it is wise to consider weaning the child away from his nebulizer while he is in the hospital.

STATUS ASTHMATICUS

Diagnosis

The child whose asthmatic attack is progressing in severity and not responding to conventional therapy (particularly epinephrine and aminophylline) must be considered to be in status asthmaticus. Every attack of asthma is potentially fatal but very rarely is; however, the chances for mortality are strikingly increased in status asthmaticus.[49,50]

The pathophysiology of status asthmaticus is an extension of the same mechanisms underlying the simple asthma attack. Airway obstruction gradually increases, leading to diminished ventilation and altered pulmonary perfusion. Eventually, if respiratory embarrassment continues, oxygen tension falls and carbon dioxide tension increases in the alveoli and perfusing blood, and respiratory acidosis develops. Then, as a result of poor fluid intake, poor caloric intake and lactic acid accumulation, metabolic acidosis supervenes. The effects of hypercapnia, hypoxia and acidosis are widespread and include central nervous system depression,[51] increased intracranial pressure, cardiac arrhythmias and hyperkalemia. Frequently superimposed is respiratory infection. Approximately one-third of all patients admitted for the treatment of status asthmaticus have pneumonia.[52] Richards and Patrick reported that infection was present in 50 per cent of their series of children who died of asthma.[53]

The clinical picture of the child in status asthmaticus is extremely grave: marked respiratory distress with heaving chest excursion, cyanosis and, later, profound alterations in the state of consciousness. O'Brien has pointed out a particularly distressing combination of signs: decreasing wheezing, decreasing breath sounds, increasing dyspnea.[54] In general, the child who has particularly loud wheezing will be adequately ventilating his chest.

Several laboratory studies should be performed after the diagnosis of status asthmaticus has been made: hemogram; arterial pH, pCO_2 and pO_2; serum electrolytes and blood urea nitrogen; chest x-ray films and electrocardiogram.

Therapy

Treatment of a child with status asthmaticus is seldom in the realm of one single physician. It requires the combined efforts of the pediatrician, allergist, pulmonary physiologist and anesthesiologist. If tracheostomy is required, the skills of the laryngologist are also required. The principles of therapy revolve around a continuation of the procedures already mentioned under treatment of the asthmatic attack. In addition, measures are taken to correct the blood gas and acid-base abnormalities and other complications such as pneumonia, pneumothorax and congestive heart failure. A few therapeutic considerations should be mentioned.

Bicarbonate Therapy

Increasing emphasis has been placed on the role of respiratory and metabolic acidosis in the creation and maintenance of the "epinephrine-fast" state.[55] It is now recommended that bicarbonate be administered intravenously for rapid correction of acidosis.[56] Here it should be emphasized that not all epinephrine-resistant asthmatics are acidotic. In fact they may be alkalotic as a result of hyperventilation.[57] Generally, before bicarbonate is administered, arterial blood gases and pH and serum electrolytes must be determined. If, however, the child is in critical condition with altered state of consciousness because of prolonged asthma, he is more than likely to be in profound acidosis and may need bicarbonate therapy immediately. In that situation it is safe to administer 2 mEq/kg body weight of sodium bicarbonate, either by slow intravenous drip or rapid intravenous administration over a 20-minute period. This dose may be repeated within thirty minutes in cases of severe acidosis. THAM (tromethamine tris aminomethane) may be used if facilities for ventilatory support are immediately available. This agent is particularly helpful when bicarbonate therapy fails to correct the acidosis initially and when the pCO_2 is extremely high.[58] If the surgeon is not familiar with the use of this potent medication, it is best to call on a colleague who is versed in this area of therapy.

Ventilatory Support

The most difficult decision in the treatment of status asthmaticus is when to abandon conventional therapeutic measures and resort to mechanical ventilation of the patient.[59] If the child's sensorium is altered, his pH drops below 7.20, his pCO_2 climbs over 65 mm Hg, and supplemental oxygen therapy does not increase his pO_2 above 50 mm Hg, endotracheal intubation and mechanical ventilation may be indicated. Bronchoscopy may be of benefit when suction of the excess mucus is indicated. It also may be of diagnostic value in suspected bronchial foreign body or in atresia.

DONT'S IN THE TREATMENT OF ASTHMA

Several rules should be followed concerning the management of the child prone to asthmatic attack:

1. Never use morphine or Demerol.
2. Never administer aminophylline or isoproterenol without first determining how much of each drug the patient has already received.
3. Never withhold steroids in the face of severe asthma, particularly in the case of the child who has required steroids previously.
4. Never prescribe drugs to which the patient has a questionable allergy.
5. Never rely upon one drug or a combination of drugs entirely in the treatment of asthma. Overdosage with epinephrine or aminophylline must be carefully avoided. Both drugs are potentially toxic.

OTHER ALLERGIC DISEASES

Allergic states other than asthma are also of concern in hospitalized children; these will be discussed briefly.

Eczema (Atopic Dermatitis)

With modern treatment, most cases of allergic eczema are easily controlled. If, however, a child with florid eczema is admitted to a hospital ward, the nursing staff will need to be cautioned on several points.

Perhaps the most important point is prevention of infection; however, since this is also the constant concern of the surgeon, the child will undoubtedly be properly shielded from this threat. Of special importance to the child's safety is the prevention of contact with children who have recently been vaccinated for smallpox. Any other source of infection, such as the infected wounds or draining sinuses of other patients, must be kept in mind as well. Even though the patient's surgical wound is healed, the eczema remains highly sensitive to infection of all types.

Another important point is the proper avoidance of known sensitizers—contactants, inhalants, foods and drugs. Finally it needs to be remembered that many of these children have had large doses of steroids, a favorite treatment of dermatologists. As noted in the care of children with asthma, these children also should receive prophylactic steroid cover.

Seldom is the degree of eczema so severe or the stay of the child in the hospital so prolonged that the surgeon need be concerned with treatment. One exception is the case of the young child with active involvement, including pruritis, weeping and crusting. In this case, scratching may easily lead to infection. Frequently this calls for restraint of the child, a measure which may also be indicated in the protection of surgical dressings, orthopedic traction devices, or intravenous needles. If the skin is dry, 1% hydro-cortisone ointment may be used topically. Ointments and creams are not suited to weeping eczema, but wet packs of Burow's solution may be indicated. It must be remembered that wet packs are difficult to keep in place in the hospitalized child; it is wise to arrange for a close attendant, perhaps the mother, to be on hand to keep them in place.

Allergic Rhinitis

Except for the fact that children with the nasal congestion and nasal discharge incident to this disease have some embarrassment of respiration, these patients should present no problem. It is a temptation to prescribe nose drops and sprays, but, except for temporary use (as during surgery), they are best avoided. The well-known tendency of such topical therapy to cause rebound congestion is a strong contraindication.

Gastrointestinal Allergy

The only problem the child with gastrointestinal allergy must face is the possibility of being fed foods to which he is allergic. Not uncommonly, medical or nursing personnel "do not believe in allergy" and will feed the child a food to which his parents say he is allergic. They would not think of administering a drug which is reported to be a cause of an allergic reaction. If the child is said to be allergic to a food, this report must be accepted at face value! It is true that in a child who has avoided a food allergen for weeks or months more or less tolerance to it may have developed. But it is a serious mistake for hospital personnel to overrule the history given by the parents and attempt to return a forbidden food to the child's diet.

Urticaria

Children with urticaria rarely need special attention. As long as the child is protected from known offenders, he should do well as a surgical patient. Again, we need to remember that many of these children have been on steroid treatment, often for long periods.

REFERENCES

1. Kantor, J.M.: Incidence of allergy in childhood. In Speer, F. (Ed.) *The Allergic Child.* New York, Hoeber, 1963, p. 36.
2. Horesh, A.J.: The role of odors and vapors in allergic disease. *J Asthma Res, 4:* 125, 1966.
3. Speer, F.: Tobacco and the nonsmoker. *Arch Environ Health (Chicago), 16:*443, 1968.
4. Girsh, L.S., Shubin, E., Dick, C., and Schulaner, F.A.: A study of the epidemiology of asthma in children in Philadelphia. *J Allerg, 39:*347, 1967.
5. Gryboski, J.D.: Gastrointestinal allergy in infants. *Pediatrics, 40:*354, 1967.

6. Idsoxe, O., Guthe, T., Wilcox, R.R., and de Weck, A.L.: Nature and extent of pencillin side-reactions, with particular reference to fatalities from anaphylactic shock. *Bull WHO, 38:*159, 1968.

7. Reisman, R.E., and Arbesman, C.E.: Systemic allergic reactions due to inhalation of penicillin. *JAMA, 203:*986, 1968.

8. Ettinger, E., and Kaye, D.: Systemic manifestations after a skin test with penicilloyl-polylysine. *New Eng J Med, 271:*1105, 1964.

9. Perlman, F.: Value of skin testing for penicillin allergy. *JAMA, 192:*644, 1968.

10. Resnik, S.S., and Shelley, W.B.: Penicilloyl-polylysine skin test, *JAMA, 196:*740, 1966.

11. Dogliotti, M.: An instance of fatal reaction to the penicillin scratch test. *Dermatologica, 136:*489, 1968.

12. Council on Drugs: *New Drugs.* Chicago, American Medical Association, 1966.

13. Israel, H.L., and Diamond, P.: Recurrent pulmonary infiltration and pleural effusion due to nitrofurantoin sensitivity. *New Eng J Med, 266:*1024, 1962.

14. Grieco, M.H.: Cross-allergenicity of the penicillins and cephalosporin. *Arch Intern Med (Chicago), 119:*141, 1967.

15. Scholand, J.F., Tennenbaum, J.I., and Cerilli, G.J.: Anaphylaxis to cephalothin in a patient allergic to penicillin. *JAMA, 206:*130, 1968.

16. Feinberg, S.M.: *Allergy in Practice.* Chicago, Year Book Publishers, 1946, p. 350.

17. Edsall, G. *et al.:* Excessive use of tetanus toxoid boosters. *JAMA, 202:*17, 1967.

18. Kaiser, K.B., Sokol, A., and Beall, G.N.: Unusual reaction to rabies vaccine. *JAMA, 193:*369, 1965.

19. Parrish, H.M.: Pitfalls in treating pit viper bits. *Med Times, 95:*809, 1967.

20. Bailey, G., Strub, R.L., Klein, L.C., and Salvaggio, J.: Dextran-induced anaphylaxis. *JAMA, 200:*889, 1967.

21. Brisman, R., Parks, L.C., and Haller, J.A., Jr.: Anaphylactic reactions associated with clinical use of dextran 70. *JAMA, 204:*824, 1968.

22. Shemin, E.R.: Anaphylactic reactions to gamma globulin. *JAMA, 203:*59, 1968.

23. Medical Research Council Working Party: Hypogammaglobulinaemia in the United Kingdom. Summary report of a Medical Research Council Working Party. *Lancet, 1:*163, 1969.

24. Falliers, C.J.: Iodotherapy appears effective in children who have asthma: Report. *JAMA, 191:*28, 1965.

25. Sheldon, J.M., Lovell, R.G., and Mathews, K.P.: *A Manual of Clinical Allergy,* 2nd ed. Philadelphia, W.B. Saunders Co., 1967, p. 137.

26. Bernstein, I., and Ausdenmoore, R.: Iatrogenic bronchospasm occurring during clinical trials of a new mucolytic agent, acetylcysteine (Mucomyst). *Dis Chest, 46:*469, 1964.

27. Fontana, V.J. *et al.:* Symposium on steroids and childhood asthma. *Clin Pediat, 7:*439, 1968.

28. Salassa, R.M., Burnett, W.A., Keating, F.R., Jr., and Sprague, R.G.: Postoperative adrenal cortical insufficiency. *JAMA, 152:*1509, 1953.

29. Slaney, G., and Brooke, B.N.: Postoperative collapse due to adrenal insufficiency. *Lancet, 272:*1167, 1957.

30. Bayliss, R.T.S.: Surgical collapse during and after corticosteroid therapy. *Brit Med J, 2:*935, 1958.

31. Braber, A.L. *et al.:* Natural history of pituitary adrenal recovery following long-term suppression with corticosteroids. *J Clin Endocr, 25:*11, 1965.

32. Robinson, B.H.B., Mattingly, D., and Cope, C.L.: Adrenal function after prolonged corticosteroid therapy. *Brit Med J, 1*:1579, 1962.

33. Livanou, T., Ferriman, D., and James, V.H.T.: Recovery of hypothalamic-pituitary-adrenal function after corticosteroid therapy. *Lancet, 293*:856, 1967.

34. Bayliss, R.L., and Steinbeck, A.S.: The adrenal response to corticotropin: Effect of ACTH on plasma adrenal steroid levels. *Brit Med J, 1*:486, 1954.

35. West, H.F.: Adrenocorticotropins and their use. *Acta Med Scand, 166* (Suppl. 352) :1, 1960.

36. Rosenblum, A.H., Rosenblum, P.: Anaphylactic reactions to adrenocorticotropic hormone in children. *J Pediat, 64*:387, 1964.

37. Howard, W. A.: Asthma. In Speer, F. (Ed.) : *The Allergic Child*. New York, Hoeber, 1963.

38. White, B., and Daeschner, C.W.: Aminophyllin (theophylline ethylenediamine) poisoning in children. *J Pediat, 49*:262, 1956.

39. Good, R.A., Vernier, R.L., and Smith, R.T.: Serious untoward reactions to therapy with cortisone and adrenocorticotropin in pediatric practice. *Pediatrics, 19*:95, 272, 1957.

40. Melby, J.C.: Adrenocorticosteroids in medical emergencies. *Med Clin N Amer, 45*: 875, 1961.

41. Siegel, S.C.: Corticosteroids and ACTH in the management of the atopic child. *Pediat Clin N Amer, 16*:287, 1969.

42. Mercer, T.T., Goddard, R.E., and Flores, R.L.: Output characteristics of several commercial nebulizers. *Ann Allerg, 23*:314, 1965.

43. Mercer, T.T., Goddard, R.F., and Flores, R.L.: Ultrasonic nebulizers, output characteristics of three. *Ann Allerg, 26*:18, 1968.

44. Keighley, J.F.: Iatrogenic asthma associated with adrenergic aerosols. *Ann Intern Med, 65*:985, 1966.

45. Van Metre, T.E.: Death in asthmatics. *Trans Amer Clin Climat Ass, 78*:58, 1966.

46. Greenberg, M.J., and Pines, A.: Pressurized aerosols in asthma. *Brit Med J, 1*:563, 1967.

47. McManis, A.G.: Adrenaline and isoprenaline: A warning. *Med J Aust, 2*:76, 1964.

48. Neder, G.A.: Death in status asthmaticus: Role of sedation. *Dis Chest, 44*:263, 1963.

49. Mosser, J.W., Peters, G.A., and Burnett, W.A.: Cause of death and pathologic findings in 304 cases of bronchial asthma. *Dis Chest, 38*:616, 1960.

50. Alexander, H.L.: Historical account of death from asthma. *J Allerg, 34*:305, 1963.

51. Kilburn, K.: Neurologic manifestations of respiratory failure. *Arch Intern Med (Chicago), 116*:409, 1965.

52. Richards, W., Siegel, S.C., Strauss, J., Leigh, M.D.: Status asthmaticus in children. *JAMA, 201*:75, 1967.

53. Richards, W., and Patrick, J.: Death from asthma in children. *Amer J Dis Child, 110*:4, 1965.

54. O'Brein, M.M., and Ferguson, M.J.: Unexpected death in bronchial asthma: A warning sign with clinicopatholgic correlation. *Ann Intern Med, 53*:1162, 1960.

55. Mithoefer, J.C., Runser, R.H., and Karetzky, M.S.: The use of sodium bicarbonate in the treatment of acute bronchial asthma. *New Eng J Med, 272*:1200, 1965.

56. Richards, W., and Siegel, S.C.: Status asthmaticus. *Pediat Clin N Amer, 16*:9, 1969.

57. Tsuchiya, Y., and Bukantz, S.C.: Studies on status asthmaticus in children: I. Capillary blood pH and pCO_2 in status asthmaticus. *J Allerg, 36*:514, 1965.

58. Strauss, J.: Tris (hydroxymethyl) aminomethane (THAM): A pediatric evaluation. *Pediatrics, 41:*667, 1968.
59. Mansmann, H.C., Marcy, J.H., and Fine, J.: Controlled ventilation with muscle paralysis in status asthmaticus. *Ann Allerg, 25:*11, 1967.

APPENDIX 1
ALLERGY QUESTIONNAIRE

Child's name _____

1. Underline any of the following active allergic conditions that the child has:
Nasal allergy, eye allergy, allergic eczema, asthma.

2. List any other significant allergic conditions _____
_____

3. Underline any of the following drugs that the child cannot take: (*This question is especially important.*)
Aspirin, penicillin, other antibiotics (list) _____,
sedatives, tranquilizers, sulfa drugs, Furadantin®, quinine, drugs for worms, ephedrine, epinephrine (Adrenalin®), asthma tablets, local anesthetics, drug for convulsions, antihistaminics, codeine, iron, atropine, cortisones, dyes used for x-ray studies, horse serum, immunizing injections (list) _____.
Other drugs: _____

4. When did the child have his last tetanus toxoid immunization? _____

5. Underline any foods the child cannot eat:
Milk, cheese, ice cream, chocolate, cola drinks, colored soda pop and drinks, egg, corn, popcorn, wheat, rice, oats, tomato, oranges, grapefruit, cinnamon, peanut butter, beans, peas. Other (include drinks, spices, nuts, etc.): _____

6. Underline any substances that cause nasal or chest symptoms:
House dust, mold, outdoor dust, feathers, animals, smoke, paint, perfumes, newspaper, room sprays, insect sprays, Lysol, floor wax, furniture polish, mothballs, alcohol, hair sprays, flowers. Other: _____

7. Underline any substances that cause rash on contact:
Soap, wool, nickel, chromates, rubber, perfumes, parabens, neomycin, clothing dyes, antibiotics, mercury, local anesthetics. (*Note:* Ignore terms unfamiliar to you.)

APPENDIX 2

ELIMINATION DIETS

Eliminate foods as circled below.

1. MILK, dried, evaporated, skim. Buttermilk, ice cream, sherbet, creamed foods, custard, cheese, cottage cheese, and milk products like Dairy Queen®. (Most patients *do not* need to avoid traces of milk in butter, bread, etc.)   _____ days

2. CHOCOLATE AND COLA DRINKS.   _____ days

3. CORN. *Corn syrup:* candies, cookies, bread, buns, canned fruit, jelly, chewing gum, peanut butter, wieners, sausage, lunch meat, ice cream. *Cornmeal:* baked goods, fish sticks, etc. *Cornstarch:* in soups, gravies, powdered sugar. Corn cereals, sweetened cereals, Fritos®, tamales, tacos, Corn Curls®, popcorn, Cracker Jacks®, hominy, grits, corn on cob, canned corn, bourbon, beer, corn flour, corn oil, corn oil margarines.

   Note: Corn must be avoided carefully! Read labels. Baked goods, canned foods, and ice cream do not require labeling by Federal law and usually contain corn.   _____ days

   [These foods do *not* contain corn: home-prepared baked foods, soups, candies, salad dressings. Granulated sugar, brown sugar, honey, spices, cottonseed oil (Wesson®, Kraft®), olive oil, Spry®, Crisco®, lard, butter, most margarines. Any vegetable but corn. Salads. Any fruit except ordinary canned fruits (dietetic OK). Prunes, dates, figs, raisins. Ham, bacon, fresh meat. Fish, chicken, turkey, shrimp, oysters. Olives, potato chips, pickles, peanuts, nuts. Pop, coffee, tea, pure sugar candy, wine, fruit juices, brandy, Scotch. Crackers, pretzels, noodles, macaroni, spaghetti, and corn-free baked goods. Egg. Milk, cheese.]

4. EGG. Baked goods (except simple breads, cookies and crackers), noodles, mayonnaise, some salad dressing, meat loaf, breaded foods, meringue, custard, French toast, divinity fudge, icings.   _____ days

5. PEA FAMILY. Beans, peas, peanuts, peanut butter, soy products, chili, honey.   _____ days

6. CITRUS FRUITS. Orange, lemon, lime, tangerine, grapefruit, and their juices. (You need *not* avoid citric acid.)   _____ days

7. TOMATO. Juice, paste, chili, catsup, soups, stews.   _____ days

8. WHEAT, OATS, RICE, BARLEY, RYE. All breads and baked goods,

flour, cake, crackers, doughnuts, cookies, waffles, pancakes, pretzels, ice cream cones, pie crust, macaroons, rolls, buns, cereals, macaroni, spaghetti, noodles, gravy. _____ days

9. CINNAMON. Spiced cakes, meats, cookies, rolls, pies, candies. Apple dishes, Dentyne®, catsup. _____ days

10. FOOD COLORS. Colored drinks (like Tang® and Koolade®), medicines, pop, bubble gum, some wieners, Jello®, popsickles. _____ days

Chapter 4

Presurgical Evaluation of the Adult Asthmatic Patient

Harold A. Lyons

THE MAJOR CONCERNS for a patient who is to undergo a surgical procedure are whether he has adequate respiratory function and whether the probability of developing postoperative respiratory complications exists. Both are related to adequacy of gas exchange arising from airway obstruction, atelectasis and pneumonia. The abnormalities can prevent operation by their severity, prolong the recovery period, or bring about severe disability or even death. The asthmatic patient already has a problem of respiratory adequacy: bronchospasm, mucosal edema and hypersecretion of mucus resulting in ventilatory insufficiency manifested clinically as wheezing, dyspnea, cough and mucoid expectoration. He may even be relatively free of these abnormalities when admitted to the hospital only to have his problem develop during surgery or in the postoperative period. The asthmatic patient thus should have a careful and complete preoperative evaluation of his status that should answer the following questions:

1. Can the operation be performed now or can it be postponed until the patient has improved ventilation?
2. Is special care needed?
3. Will special anesthetic management be required?
4. Should the planned operation be modified or substituted for another procedure?
5. What should be the postoperative management?
6. Which is the greater risk, delay of surgery or proceeding with added risk due to suboptimal pulmonary function?

The preoperative evaluation of the adult asthmatic patient includes all the procedures used for any patient who has pulmonary disease. Asthma may be associated with significant abnormalities of respiratory function ranging from minimal or absent to maximal impairment, since it is a disease with changing degrees of severity. Thus the evaluation should include the current state of the patient and attempt to delineate what an increase of symptoms might indicate.

The history is important for ensuring diagnosis is correct and determining whether an associated respiratory disease is or is not present. Effort

Note: This work was supported in part by U.S. Public Health Service Grant 5R01 HE 11932-02.

should be made to rule out those diseases which simulate asthma clinically, for example, left ventricular failure, chronic obstructive airway disease (emphysema) with alveolar destruction and chronic bronchitis.[1]

PULMONARY FUNCTION TESTS

The most useful pulmonary function tests for preoperative evaluation of any patient are tests of the dynamic ventilation and the arterial blood gases.[2,3]

Spirometry is readily available and is the time-honored method for testing ventilation. With it, the vital capacity, the total excursion of the lungs, and the velocity of air movement in and out of the chest can be measured. The vital capacity, if volume is delivered slowly enough, may approach normal values in the asthmatic patient. The degree of obstruction is characterized and quantitated by the slowed maximal airflow rates. The reversibility of airway obstruction is determined by the amount of improvement in maximal airflow rates after administration of a bronchodilator.[4,5]

Tests of Expiratory Airway Obstruction

Methods for demonstrating and quantitating the expiratory airway obstruction are numerous.* Various indices employ the spirographic tracing and are useful and simple to obtain, including the timed vital capacity and maximal expiratory flow rate. The maximal voluntary ventilation (previously called the maximal breathing capacity) is another test which ordinarily employs spirometry. Despite considerable pro-and-con discussion of different methods, the reported findings are similar. Slowing of expiratory flow indicates with reasonable sensitivity and specificity that an excessive operative risk exists, especially when this abnormality is unrelieved by bronchodilators (Fig. 4-1 and 4-2). In patients with airway obstruction, the rate of complications following surgery is high. The risk of complication developing during or after surgery is seven to nine times greater than in patients with normal airflow dynamics.[2-8]

It has been observed that in asthma, when the forced expiratory volume per second (FEV_1) is 15 to 20 percent, carbon dioxide retention is present[9] (Table 4-1).

If slowing of the expiratory flow is observed, the surgery should be postponed if possible until respiration is improved with therapy. If surgery cannot be delayed, then bronchodilator therapy should be administered and the tests repeated in an attempt to obtain maximal relief of airway obstruction. One should then be aware that during the surgical procedure anesthetic difficulties may arise.

---

*See Appendix to this chapter.

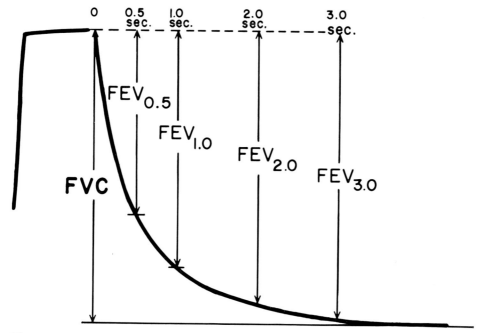

Figure 4-1. The spirographic trace of forced exhalation. Note that normally 75 percent of the full-forced vital capacity can be delivered in one second.

If the vital capacity is reduced, a restrictve pulmonary disease may be augmenting the greater operative risk of obstructive defect. One may also question the diagnosis if the amount of obstruction present cannot fully explain the reduced vital capacity.

Test of Arterial Blood Gases

No pulmonary function test other than determination of the arterial blood gases indicates the adequacy of gas exchange in the lungs. This evaluation should be routine and within the capability of every hospital. In asthma, the arterial blood gases may be normal, and this finding would argue for less risk from surgery. However, the most significant finding during an attack of asthma is hypoxemia. The uneven distribution of inspired air resulting from airway obstruction produces regional areas of hypoventilation from which the blood extracts little oxygen. These areas are perfused with blood to a better degree than they are ventilated and thus have a low ventilation-perfusion ratio. When they are numerous or extensive, the resulting arterial hypoxemia is always serious and will greatly add to surgical risk and complications. It may actually contraindicate surgery.

Associated with hypoxemia in asthma is an increased physiological dead space in the lung. This dead space is produced by other regional areas with-

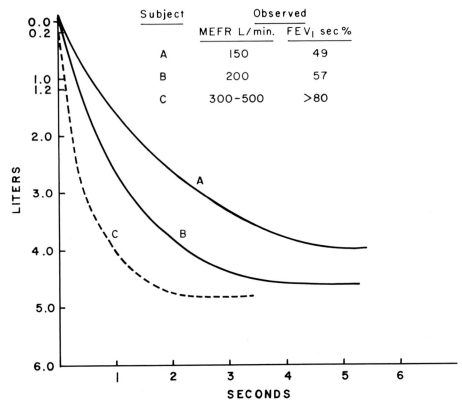

Figure 4-2. The spirographic tracings of an asthmatic subject. *A:* Note the prolonged time necessary for the airflow to be delivered. *B:* After the administration of a bronchodilator there is considerable improvement in the flow rate. Comparison of both flow rates at a similar volume of inflation is useful in distinguishing between the effect of the bronchodilator on lung volume and its effect on improving airway resistance. It must be remembered that the spirographic record cannot demonstrate the resistance to airflow present in small airways (see text). *C:* The spirographic tracing of a normal subject for comparison.

in the lung which are overventilated in relation to their perfusion to such a degree that arterial carbon dioxide is reduced and respiratory alkalosis results. When the airway obstruction is advanced, however, more generalized

TABLE 4-1

PERFUSION VALUES OF CARBON DIOXIDE IN ARTERIAL BLOOD IN RELATION TO PERCENT OF ONE SECOND FORCED EXPIRATORY VOLUMES

Range % of FEV_1	$PaCO_2$ mm Hg
50 to 60	25 ± 3.8
40 to 20	33 ± 4.4
20 to <15	39 ± 9.0

hypoventilation occurs, and ensuing arterial carbon dioxide retention and respiratory acidosis are observed.[9,10] This condition does not permit surgery and must be vigorously treated.

Correlating spirometric and blood-gas tension measurements indicate that blood-gas disturbance is related closely to the degree of reduction in the forced vital capacity (FVC) even more than the degree of obstructive defect, as measured by the forced expiratory volume in one second (FEV_1). Hypoxemia is worse as increase of airway resistance occurs and overinflation of the lung increases. Woolcott and Read[13] place importance on changes in functional residual capacity (FRC) and total lung capacity (TLC) for assessment of the severity of obstruction. These lung volume and ventilatory changes relate to the disturbances in blood-gas tensions in asthma. The relationship between the tension of arterial carbon dioxide ($PaCO_2$) and the ventilatory changes are different from those involved in the production of hypoxemia. Early in the attack there is a reduced $PaCO_2$ hypocapnia due to alveolar hyperventilation often with an accompanying respiratory alkalosis. These low levels of $PaCO_2$ usually persist even when airway obstruction and hypoxemia increase in severity. A stage is eventually reached when lung hyperinflation is so great that the required elastic work of breathing is beyond the capacity of the patient. Alveolar ventilation decreases, and $PaCO_2$ rises, reaching normal levels as the condition is worsening, and finally entering a critical state of respiratory acidosis[8-10] (Table 4-1). Failure to appreciate the relation of the occurrence of a normal $PaCO_2$ with the spirometric data of severe increase in airway resistance may result in false assurance regarding the clinical state of the patient.

Other Pulmonary Function Tests

The other pulmonary function tests such as nitrogen-washout studies, compliance, and diffusion-capacity determinations ordinarily contribute little in the evaluation of a patient for surgery. However, they are useful in assessing obstruction of the small airways, a condition not disclosed by any measurement made at the mouth.[11,12] This undetected abnormality was called to our attention during clinical experience with asthmatic patients when we observed patients who had been satisfactorily treated and relieved of respiratory distress die unexpectedly. In each instance, necropsy disclosed small airway passages plugged with mucus and casts. In patients with asthma, changes in clinical severity may be accompanied by changes in total lung capacity rather than in FEV_1.[12] Other ways this abnormality can be recognized are by detection of persistent arterial hypoxemia and of uneven ventilation by, for example, the washout of nitrogen from the lung by breathing 100% oxygen, or by determination of pulmonary compliance at various respiratory frequencies and observing frequency-dependent compliance. Normally, pulmonary compliance remains unchanged up to frequencies of 40 to 60 per minute.[14] For a complete evaluation of the asthmatic patient, either of these tests should be included to unmask small airway obstruction, even though normal values may be observed with the usually employed tests.

SUMMARY

The presurgical evaluation of the asthmatic patient should include (a) certainty of the diagnosis; (b) tests for expiratory rate of airflow, and if abnormal, whether these tests can be improved by a bronchodilator; (c) determinations of arterial blood gases; (d) tests to rule out small airway obstruction, and (e) measurement of diffusion capacity to rule out destructive lesions in the lung. If these tests are normal, the patient can be judged able to undergo surgery; if abnormal and surgery can be delayed, then therapy should be administered until optimal improvement is attained and the tests are near or within the normal range. If surgery must not be delayed, then the abnormal function shown by the tests must be improved as far as possible, special attention given to ventilation and oxygenation during surgery, and preparations for postoperative ventilatory care made. Hypoxemia is frequent during an attack of asthma and increases surgical risk. All abnormalities as revealed by the pulmonary function tests should be corrected or minimized before a surgical procedure is undertaken.

The test for diffusion capacity of the lung is useful for determination of the correctness of the diagnosis. In asthma, a normal diffusion capacity of the lung is expected. However, if it is low, the suspicion of destructive lesions arises, and in all probability chronic bronchitis or emphysema is present. This test is advised for inclusion in all preoperative evaluations.

Spirometry alone is not sufficient for preoperative evaluation of pulmonary function; and if the criteria of spirometry alone are used, the resulting disability will be greater than expected.

REFERENCES

1. Williams, D.A.: Classification of asthma and its therapeutic implications: The nature of asthma: Report of a symposium held at King Edward VII Hospital, Midhurst, Sussex, England, 1964, Section 5, pp. 85-94.
2. Swenson, E.W., Stallberg-Stenhagen, S., and Beck, M.: Arterial oxygen, carbon dioxide and pH levels in patients undergoing pulmonary resection. *J Thorac Cardiovasc Surg, 42:*179, 1961.
3. Stein, M., Kosta, G.M., Simon, M., and Frank, N.A.: Pulmonary evaluation of surgical patients. *JAMA, 191:*765, 1962.
4. Comroe, J.H., Jr. *et al.: The Lung, Clinical Physiology and Pulmonary Function Tests,* 2nd ed., Chicago, Year Book Medical Publishers, 1962.
5. Malmberg, R., Dotton, O., Berglund, E., Simonsson, B.G., and Bergh, H.P.: Preoperative spirometry in thoracic surgery. *Acta Anesth Scand, 9:*57, 1965.

6. Gaensler, E.A. *et al.:* The role of pulmonary insufficiency in mortality and invalidism following surgery for pulmonary tuberculosis. *J Thorac Surg, 29:*163, 1965.

7. Mittman, C.: Assessment of operative risk in thoracic surgery. *Amer Rev Resp Dis, 84:* 197, 1961.

8. Rothfield, E.L. *et al.:* Pulmonary function testing in geriatric surgical patients. *Dis Chest, 41:*85, 1962.

9. McFadden, E. R., Jr., and Lyons, H.A.: Arterial blood gas tension in asthma, *New Eng J Med, 278:*1027, 1968.

10. Tai, E., and Read, J.: Blood gas tensions in asthma. *Lancet, 1:*644, 1967.

11. Macklem, P., and Mead, J.: Resistance of central and peripheral airways measured by a retrograde catheter. *J Appl Physiol, 22:*395, 1967.

12. McFadden, E.R., Jr. and Lyons, H.A.: Airway resistance and uneven ventilation in bronchial asthma. *J Appl Physiol, 25:*365, 1968.

13. Woolcock, A.J., and Read, J.: Lung volumes in exacerbations of asthma. *Amer J Med, 41:*259, 1966.

14. Woolcock, A.J., McRae, J., Morris, J.G., and Read, J.: Abnormal pulmonary blood flow distribution in bronchial asthma. *Aust Ann Med, 15:*196, 1967.

15. Mills, R.J., Cumming, G., and Harris, P.: Frequency dependent compliance at different levels of inspiration in normal subjects. *J Appl Physiol, 18:*1061, 1963.

APPENDIX

A brief description of the technics for the performance of the pulmonary function tests discussed in the text is presented here. For a fuller discussion and more detailed features, consult references at the end of this appendix.

Vital Capacity

The vital capacity (VC) is the oldest lung function test used. It describes accurately the static volume or maximal displacement of the respiratory pump. The test is performed by having the subject take a full inspiration and then exhale completely. This exhalation can be recorded on a spirometer bell with kymograph. Approximately 25 ml/cm of height for men and 20 ml/cm for women are normal values.

Dynamic Ventilatory Tests

A number of simple tests measure not only the total vital capacity but also some parameter reflecting the rate of the flow during expiration.

The *timed vital capacity* uses the rapid delivery of a vital capacity and measures the percent which can be delivered in one second (see Fig. 4-1). Normally 75 percent of the vital capacity can be expelled within one second and over 90 percent within three seconds.

The *maximal midexpiratory flow rate* is the volume exhaled after a forced expiration and is that volume in the midportion of the vital capacity. The *maximal expiratory flow rate* is determined by measuring the expired first liter of volume in relation to time, from 200 to 1200 ml. The first 200 ml is disregarded to account for hesitation of the patient in delivering full flow and for the inertia of the spirometer bell. Normal predicted values are available in the nomograms of Kory, Callahan, Boren and Syner[5] (Fig. 4-1). The normal values range between 400 to 600 liters per minute.

These tests require a low-resistance spirometer system and a fast-moving spirograph. As can be observed, all of these determinations can be derived from the single forced expiration. These tests should always be repeated several times and again after use of a bronchodilator. Improvement of an obstructive pattern greater than 10 to 15 percent is considered indicative of reversible airway obstruction.

All of these dynamic ventilatory tests distinguish obstructive airway disease. Although there are various reasons given for the superiority of one test over the other regarding the portions of the curve for use in calculation, actually from a practical viewpoint all provide essentially the same information.

Pulmonary Diffusion Capacity

The measurements of pulmonary diffusing capacity can be made by several techniques. All frequently used tests employ carbon monoxide as the test gas. The single breath method of Ogilvie and associates is the easiest to perform. The reader who plans to employ this test is advised to consult the references.[1-3]

The test consists of the patient inhaling a known concentration of carbon monoxide (usually 0.5%) from residual volume of the lung to full capacity, holding the breath for ten seconds and then slowly exhaling, during which the operator collects the sample of the expired gas after the dead space is cleared. The difference between the initial and final concentration of C0 represents the uptake of carbon monoxide. Since the decrease in concentration is an exponential function, the calculation of diffusion of carbon monoxide in the lung (DL_{C0}) is made by the following equation:

$$DL_{C0} = \frac{ml\ C0\ transferred\ from\ gas\ to\ blood/min}{mean\ alveolar\ C0\ pressure\ -\ mean\ pulmonary\ capillary\ C0\ pressure}$$

The factor involved in diffusing capacity are (a) surface area for diffusing; (b) distance for diffusion; (c) characteristics of the tissue; and (d) solubility of the gas.

Evenness of Distribution

The evenness of the distribution of inspired air is a test requiring complex apparatus: a spirometer with recording kymograph, a valve box and recording meter for inert gas concentration. In principle, the test is performed by either washing out nitrogen from the lungs by breathing 100% oxygen or by breathing in a known concentration of an inert gas until the lung comes into equilibrium with the spirometer in a closed system. With the first method less than 2% nitrogen is present in the exhaled breath after three minutes for normal subjects. The same time interval is required for the inert gas to reach equilibrium. Helium is usually used in the closed system method. A longer time is associated with unevenness of distribution of inspired air due either to increased airflow resistance or compliances randomly distributed throughout the lung (Fig. 4-3).

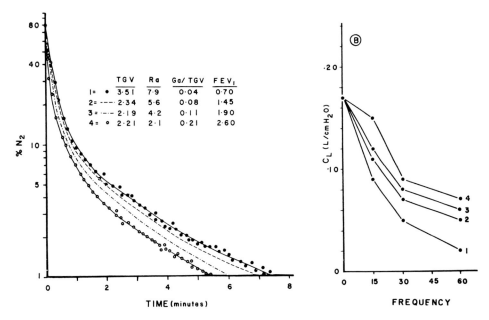

		TGV	Ra	Ga/ TGV	FEV$_1$
1=	•	3·51	7·9	0·04	0·70
2=	----	2·34	5·6	0·08	1·45
3=	-·---	2·19	4·2	0·11	1·90
4=	○	2·21	2·1	0·21	2·60

Figure 4-3. The observed nitrogen washout study on an asthmatic patient (*left*), after improvement of his airway resistance (Ra) with bronchodilator therapy, and the recorded compliance measurements (C$_L$) made at various respiratory frequencies per minute done at the same time as the nitrogen washout study (*right*). The thoracic gas volume (TGV), conductance-thoracic gas ratio (Ga/TGV) and forced expiratory volume in one second (FEV$_{1.0}$) are also shown. There is no appreciable changes in either of these measurements even though airway resistance have reached the normal value (2.1 L/cm H$_2$O/sec), indicating the persistence of lower airway abnormality for airflow.

Arterial Blood Gases

Arterial blood sampling should be regarded as an essential procedure in management of respiratory problems.[2,4] The sample is taken in a prepared sealed syringe which contains heparin. It should not contain air bubbles and should be sealed. The pH, pCO$_2$ and pO$_2$ are determined by the blood-gas electrode systems (Instrumentation Laboratory Associates, or Radiometer).

It is best to have the analysis performed within minutes; if there is a delay, the sample should be placed in ice until analyzed. The normal value for arterial blood are PaO_2 90 to 100 mm Hg, $PaCO_2$ 38 to 42 mm Hg, pH 7.38 to 7.42.

References

1. Comroe, J.H., Jr. *et al.: The Lung. Clinical Physiology and Pulmonary Function Tests,* 3rd ed. Chicago, Year Book Publishers, 1965.
2. Cotes, J.E.: *Lung Function Assessment and Application in Medicine,* 2nd ed. Philadelphia, F.A. Davis, 1968.
3. Kory, R.C.: Screening techniques for early pulmonary function impairment. *Arch Environ Health (Chicago), 6:*155, 1963.
4. Bartels, H. *et al.: Methods in Pulmonary Physiology.* Translated by J.M. Workman. New York, Harper, 1963.
5. Kory, R., Callahan, R., Boren, H., and Syner, G.: Cooperative study of spirometry: Veterans Administration Hospital. *Amer J Med, 30:*243, 1961.

Chapter 5

Presurgical Evaluation of the Allergic Child

Roy F. Goddard

M ANY DETAILS of medical care arise when an allergic child is admitted to the hospital for surgery, whether it be major or minor. First, a definite diagnosis must have been previously established, and the chronicity and severity of the condition must be analyzed as well as the patient's ability to withstand the procedure. In addition to the simple, ordinary laboratory tests, a thorough evaluation of the child's pulmonary function is imperative. His response to medical management is of considerable importance; he must reach his maximal, or at least optimal, physical and mental condition before surgery is attempted.

THE DIAGNOSIS

The first important consideration in the presurgical evaluation of an allergic child is establishing the diagnosis. As emphasized by Speer and Shira (Chap. 3), the child's condition should be identified in a simple category such as eczema, allergic rhinitis, gastrointestinal allergy, urticaria, or a combination of allergies. (The asthmatic child is in another classification or category of a more serious nature.) Of great importance is his ability to adjust to stress. If he is unable to adapt to the usual, and some of the unusual, stresses of everyday life, he must receive special attention and management to prepare for the approaching surgery and procedures. When the child is acutely ill, obviously surgery must be delayed until his illness can be controlled or he is better prepared. If the patient is suffering from a subacute or chronic disease other than allergy, then this also should be under control. Cystic fibrosis may occur in an allergic child and present complications, as do pneumonia and other bacterial infections. If such complications are too numerous or marked in severity, surgery may be contraindicated.

Obtaining a thorough history of the child scheduled for surgery, including his allergies, clinical response and management together with the present status of the patient, is mandatory. One must have or gain an insight into the future of the patient to decide whether he will benefit from surgery or suffer complications from the procedure.

The physical findings are contributory to the presurgical evaluation. Obviously, the general appearance of the patient, his nutrition, color, hydration, presence or absence of anemia, and signs of obstructive lung disease (clubbing, cyanosis, dyspnea) are all of significance. A productive cough and involvement of the child's upper respiratory passageways may present problems. The patient may demonstrate an allergic facies (swollen eyelids, widening and flaring of the alae nasi, cyanosis of the lips), with involvement of the turbinates (Fig. 5-1). Nasal polyps or enlarged adenoids may cause upper airway obstruction.

The configuration of the thorax may indicate severity of the disease process. An increase in the anteroposterior (AP) diameter and an increase in the spaces between the ribs denote chronic hyperaeration and loss of compliance, usually with decreased excursion and limitation of movement of the chest (Fig. 5-2), thus indicating restrictive impairment.[1]

Congenital anomalies that influence the ventilation of the lung, such as small upper respiratory passageways, laryngeal webs, and those conditions that may cause obstruction during surgery should be noted. Further evidence of airway obstruction may be found in an increased respiratory rate, shallow depth of breathing, flaring of the alae nasi, and costal, subcostal or supercostal or superclavicular retractions. Dyspnea, at rest or on exertion, and the patient's exercise tolerance or stress should be evaluated. The size

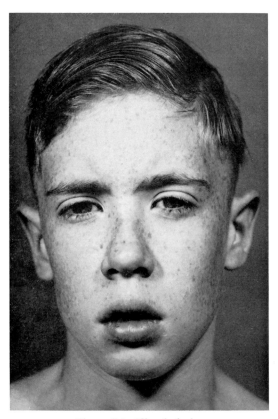

Figure 5-1. Allergic facies.

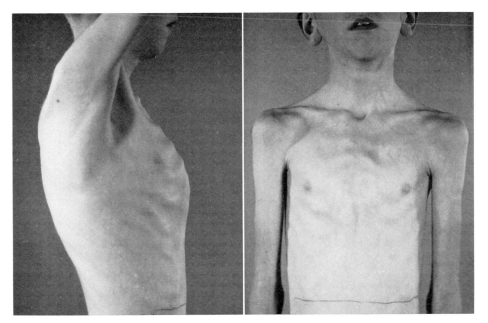

Figure 5-2. The pre-emphysematous chest of the severly intractable asthmatic child (*left,* lateral view; *right,* frontal) .

of the heart, its position, rate, rhythm and presence of any murmurs can be significant. Other body systems too may be involved in respiratory obstruction: the liver may be enlarged or there may be increased sweating or other evidence of increased metabolism.

Of considerable importance to establishing a diagnosis are laboratory studies. The simple, general laboratory tests include a complete blood count, urinalysis, nasal smear for eosinophils, throat or sputum culture (with sensitivities if indicated) , total proteins, and x-ray films of the chest. More complex procedures define specific systems and the severity of their involvement.[2,3] These studies include sweat electrolytes (by iontophoresis method) , immunoglobulins (by electrophoresis), serum electrolytes and pH, pCO_2, and bicarbonate; liver function and pulmonary and other function studies (if indicated) .

X-ray films of the chest (AP and lateral) should be taken on deep inspiration to show configuration of the chest, amount of aeration, and cardiopulmonary relationships. Frequently there may be hyperaeration and a barrelchest with increased AP diameter, increased rib spaces, flattening of the diaphragm, and in some instances, cor pulmonale or pulmonary hypertension, or both (Fig. 5-3) . Sinus x-ray films and sagittal or lateral views of the neck to indicate large lymphoid or adenoid tissue are also important in determining upper airway obstruction.

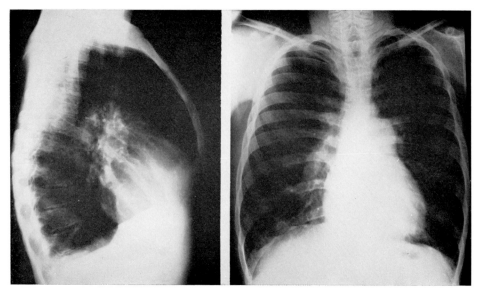

Figure 5-3. Chest roentgoengram of a teen-ager with severe asthma and emphysema (*left,* lateral view; *right,* frontal) .

EVALUATION OF PULMONARY FUNCTION

One of the most important facets in the presurgical evaluation of the allergic child is the evaluation of his pulmonary function on both a quantitative and qualitative basis. There has been considerable discussion on the difficulties of studying pulmonary function in children, but these have been surmounted in many instances by utilizing extreme patience in studying little patients. One can study satisfactorily these values in children four years of age and older, and in some instances two years of age, and by specialized mechanics of breathing studies in even younger age groups and infancy. Obviously, good cooperation from the patient is essential to the successful completion of a pulmonary function test, as is the repetitiveness of the test to determine its reliability and interpretation. The severity of the disease process may influence the succcess of the test and completion in its entirety. In some instances, a test may not be achieved if a child is severely or acutely ill at the time a pulmonary function test is scheduled. Therefore, one should try to have the patient in optimal condition before an evaluation of his pulmonary physiology is undertaken.

In general, one can study airflow, lung volumes, general respiratory competence and intrapulmonary distribution and mixing.[4] Restriction of the lungs may be determined by measuring vital capacity, total capacity and maximal breathing capacity, while obstruction may be studied by measurements of timed vital capacity or maximal midexpiratory flow and the flow-volume curve. In addition, in the flow-volume studies, we can determine the effect of a bronchodilator, ether, or other medications or anesthetics by simple repetitive studies following inhalation of the particular medication.

Values for intrapulmonary mixing and distribution can be obtained from nitrogen clearance studies[5] and pulmonary diffusion by the steady state carbon monoxide method of Bates and associates.[6] Blood gas determinations can be made from arterialized capillary blood studies, using micromethods for pH, pCO_2, and $BACO^3$ according to Astrup,[7] in younger children, or by direct arterial punctures in older children.

Resting tidal ventilation and minute volume are recorded with the child breathing air on a 9-liter Collins Spirometer in a closed circuit, nonrebreathing system to establish the end tidal level (Fig. 5-4). Such a test may easily

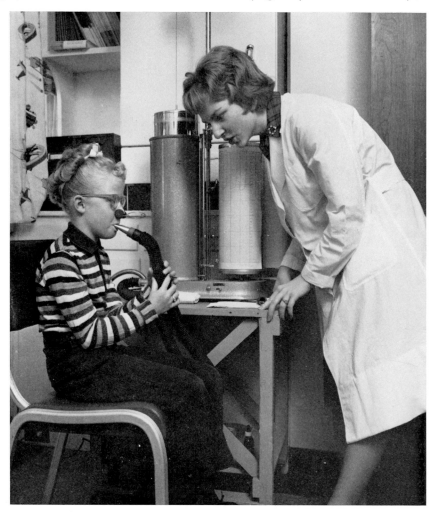

Figure 5-4. Spirometric study using 9-liter Collins Spirometer.

be used in most hospitals and laboratories and furnishes a spirometric tracing, which indicates the tidal volume exchange and expiratory capacities. By increasing the slow speed of 3.2 cm/min to a faster 16.0 cm/min, one may obtain a timed vital capacity (Fig. 5-5).

Following spirometric measurements, pulmonary mixing and distribution may be studied at the same time that functional residual capacity (FRC) is measured by using an open circuit system similar to that described by Darling (Fig. 5-6).[8] While the child breathes oxygen, the expired air is collected in a large spirometer. From the amount of nitrogen in the expired volume, the FRC may be calculated. At the same time, the fall in the fraction of nitrogen in the expired gases is indicated, breath by breath, by the nitrogen meter until the end tidal nitrogen concentration has dropped below 1% (Fig. 5-7). This nitrogen fraction is recorded simultaneously with the expired volume. The ratio of the volume of ventilation required to clear the lungs of nitrogen to the volume being ventilated (the FRC) is an indi-

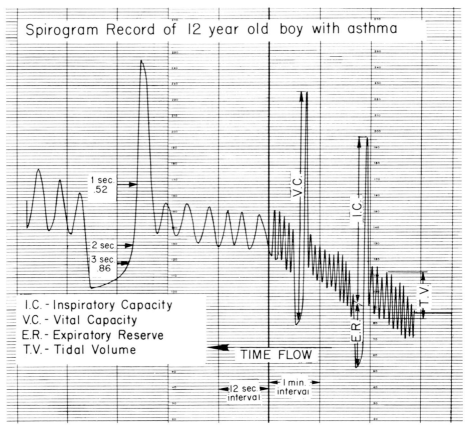

Figure 5-5. Spirogram record demonstrating change in spirometric speeds for timed vital capacity maneuver (9-liter Collins Spirometer).

cation of the efficiency of ventilation and is called the *clearance equivalent.* The residual volume is obtained by subtracting the expiratory reserve, measured on the conventional spirometer tracing, from the FRC. The total capacity is the sum of the residual volume and the vital capacity.

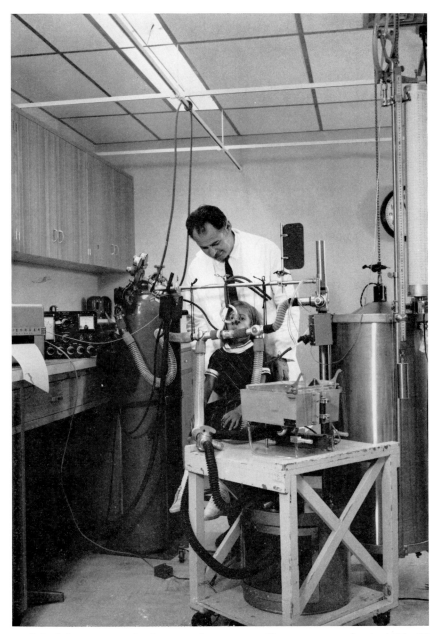

Figure 5-6. Spirometry and nitrogen clearance studies on a preschool child (Krogh Spirometer and Med-Science Nitralyzer) .

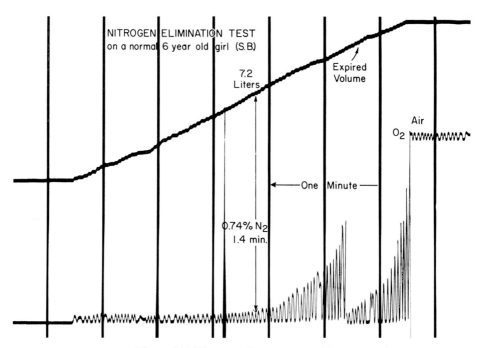

Figure 5-7. Nitrogen clearance recording.

In the case of preschool children (2 to 6 years), a 1.5 liter Krogh Spirometer is used to record resting ventilation, and whenever possible, the vital capacity. When this study is done, a combined tracing of the spirometric and diffusing studies can be recorded on the Visicorder as seen in Figure 5-8, showing a comparative spirometric and mixing study of a normal 4-year-old and a 4-year-old asthmatic child who demonstrates obstructive, restrictive and mixing impairment.

From these pulmonary function studies, one must then determine the type of ventilatory impairment present—restrictive or obstructive, or a combination of the two. In addition, the severity of impairment can be evaluated from pulmonary function tests, response to certain medications, and therapy which will allow the patient to approach a more normal condition simulating an optimal or safe condition for surgery. The values obtained in the subject under study can be compared with the normal tables and regression curves for that age group (Fig. 5-9) in order to determine if the child has sufficient total capacity to undergo surgery.

To study the individual respiratory cycle or gas exchange during one complete respiratory cycle, accurate flow-volume studied can be obtained by the use of a Sanborn X-Y Recorder in conjunction with a wedge spirometer (Fig. 5-10). The recording obtained from this study is in the form of a flow-

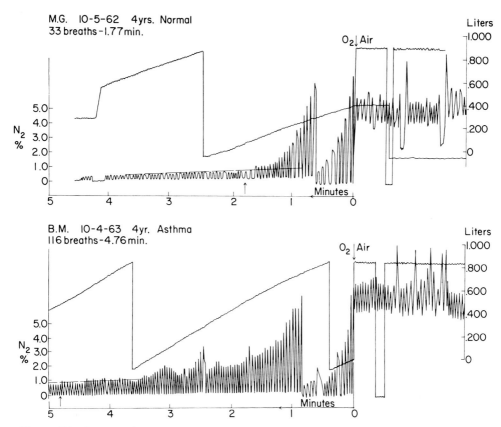

Figure 5-8. Comparative spirometric and mixing studies of normal 4-year-old and 4-year-old asthmatic child who demonstrates obstructive, restrictive and mixing impairment.

volume loop, the X axis being the volume axis and the Y axis representing the rate of flow. Normal and maximal forced inspiratory and expiratory efforts can be achieved. The most useful values obtained are the peak expiratory flow, the maximal midexpiratory flow, the midinspiratory flow, and the forced vital capacity. This flow-volume single breath method has replaced the conventional maximal breathing capacity studies in our laboratory for the evaluation of obstructive impairment in children. Significantly, from such flow-volume studies, one can compare not only the flow-volume measurements of the patient and progress of his condition, but also the effects of bronchodilating agents or other therapeutic medications, or the use of mechanical ventilating devices, nebulizers and breathing exercises on the patient (Fig. 5-11, Top and Bottom).[4]

Pulmonary diffusing capacity can be measured satisfactorily in children over six years of age by the steady state carbon monoxide method. Arterial punctures are technically difficult to perform on small children and are un-

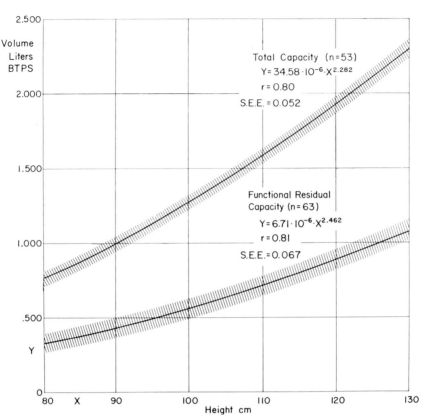

Figure 5-9. Regression curve correlating total capacity and functional residual capacity with height in preschool children.

comfortable for them. However, arterialized capillary blood studies employing the previously mentioned micromethods of Astrup,[7] are frequently of value in selected groups. These capillary samples are obtained after adequate arterialization by local application of heat to a finger, earlobe, or heel. Simultaneous collection of expired air permits estimation of alveolar ventilation and of the alveolar-arterial oxygen gradient.

Rarely are physical competence studies required in presurgical evaluation, but these can be done by measuring the oxygen consumption during maximal physical exertion at the limits of an individual's aerobic capacity. A simple graded exercise test is performed on a bicycle ergometer at work loads appropriate to the test subject's size. Modified competence studies can be utilized in studying the more severely afflicted patient.

MEDICAL MANAGEMENT

It is most important that the physician try to improve or correct a reversible condition. If this can be done by intensive medical therapy, every

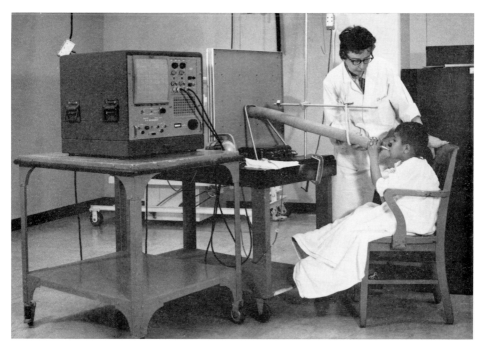

Figure 5-10. Flow-volume study employing wedge spirometer and Sanborn X-Y Recorder.

effort should be made to improve to the maximal or optimal level with clinical subjective observations and objective pulmonary function tests utilized as guides to achievement. Such management obviously involves all modalities of treatment and control of the allergic manifestations in all segments of the respiratory tract. Bronchial hygiene, including the administration of bronchodilators, wetting agents or detergents, enzymes, and the use of intermittent positive pressure breathing, may offer significant improvement in these areas. If alteration of respiratory obstruction is not sufficient, one may have to cancel or postpone the anticipated surgery until such optimal alteration can be achieved.

Summary of Presurgical Evaluation

At this point, one should summarize the presurgical evaluation of the allergic child by determining the diagnosis, its severity, the possibility of reversibility, whether the patient can tolerate anesthesia and the proposed surgical procedure, and what postoperative complications might arise. The future expected course of the child, plotted not only from the history but from the physical findings and objective laboratory data, may require careful scrutinizing. All these factors may suggest that surgery be postponed. Particularly important is evaluation of the response to bronchospasm, de-

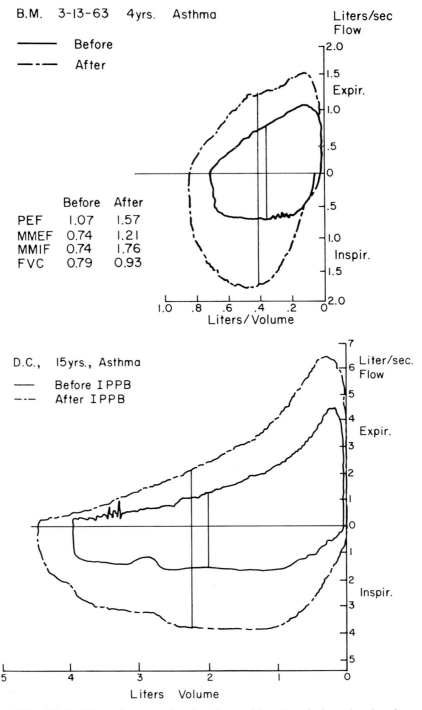

Figure 5-11. (*Top*) Flow-volume study of a 4-year-old asthmatic boy showing improvement after bronchodilator therapy with intermittent positive pressure breathing. (*Bottom*) Flow-volume study of a 15-year-old boy with chronic asthma, demonstrating typical. adult emphysematous type of respiratory loop, from the same type of therapy.

creased vital capacity or restrictive impairment that may occur at rest, under stress, or during surgical procedures and when further obstruction of ventilation may be expected. In addition, one must determine what medication (s) , including the type of anesthetic to be used, may safely be given to the child (Chap. 8) .

Recommendations Relative to Surgery in the Allergic Child

With such a summary of the presurgical evaluation at hand, the pediatrician may inform the surgeon of any reservations he has with respect to the procedure and the selection of an anesthetic. In addition, he should give some estimation of the duration of the procedure the child can tolerate, insist that airway provisions are available, together with adequate ventilatory measures, provisions for blood loss and other complications which might arise during the period of surgery. The pediatrician should recommend to the surgeon the type of immediate postoperative care that is indicated and also the long-term care for the child in whom complications may develop. In some cases contraindications to surgery may be absolute and prevent surgery while in others they may indicate that only after therapy will the child be able to tolerate surgery.

Summary of Important Considerations in Surgery of the Allergic Child

The important considerations in preparing an allergic child for surgery have been reviewed. One must ask these imperative questions: What is the diagnosis? Can a specific diagnosis be offered? What is the severity of the condition; the chronicity, the expected course without surgery? What can be expected with surgery? What laboratory tests are indicated? Have these tests been done and what results have been obtained from them? In centers where pulmonary function studies are available, it is essential that complete volumetric, mixing and distribution, diffusion, and flow-volume studies be done. In other hospitals and laboratories where these more sophisticated evaluations are not available, simple screening tests can be performed with such instruments as the Wright Peak Flow Meter (Fig. 5-12) for determination of obstruction and the Collins Spirometer (Fig. 5-4) or the Jones Pulmonor (Fig. 5-13, 5-14) for determinations of vital capacity, timed vital capacity, and maximal midexpiratory flow. The postoperative follow-up with reference to immediate or prolonged care is imperative, as are repeated, periodic pulmonary function tests and observation of the course and success of the surgery.

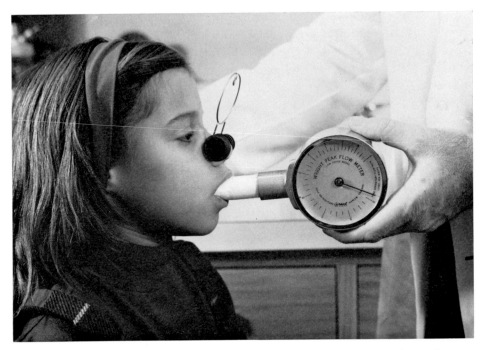

Figure 5-12. Wright Peak Flow Meter (pediatric) for studying airway obstruction.

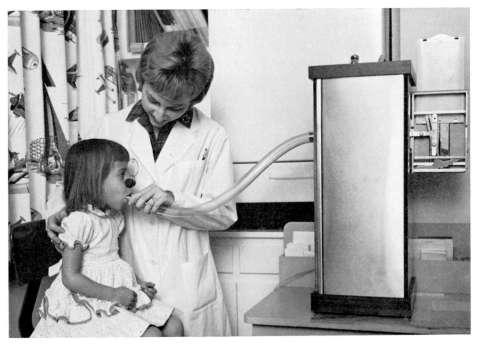

Figure 5-13. Jones Pulmonor for use in office or hospital.

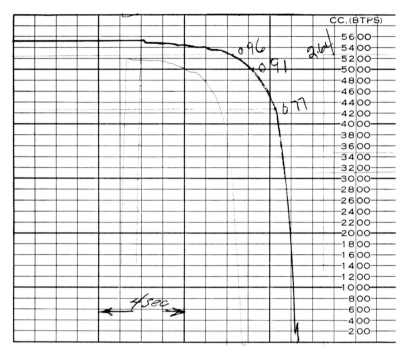

Figure 5-14. Tracing with Jones Pulmonor.

REFERENCES

1. Goddard, R.F.: Pre-emphysema in children, its recognition and treatment. *Ann Allerg, 19:*1125, 1961.
2. Goddard, R.F.: The physiologic-pathologic-clinical relationships of chronic pulmonary diseases in children. *J Child Asthma Res Inst & Hosp, 1:*31, 1961.
3. Goddard, R.F.: *Clinical Cardiopulmonary Physiology.* New York, Grune & Stratton, 1960, p. 724.
4. Goddard, R.F., and Luft, U.C.: *Pulmonary Function Tests for Infants and Children.* Albuquerque, (New Mexico), Lovelace Foundation for Medical Education and Research, 1969.
5. Luft, U.C., Roorbach, E.H., and MacQuigg, R.E.: Pulmonary nitrogen clearance as a criterion of ventilatory efficiency. *Amer Rev Tuberc, 72:*465, 1955.
6. Bates, D.V., Wolf, C.R., and Paul, G.I.: A report on the first two stages of the coordinated study of chronic bronchitis in the department of veterans affairs, Canada. *Med Serv J Canada, 18:*211, 1962.
7. Astrup, P.: A simple electrometric technique for the determination of carbon dioxide tension in blood and plasma, total content of carbon dioxide in plasma, and bicarbonate content in "separated" plasma at a fixed carbon dioxide tension (40 mm Hg). *Scand J Clin Lab Invest, 8:*33, 1956.
8. Darling, R.C., Cournand, A., and Richards, D.W., Jr.: Studies on the intrapulmonary mixture of gases. III. An open circuit method for measuring residual air. *J Clin Invest, 19:*609, 1940.

Chapter 6

Surgical Treatment of Asthma in Adults

Brian Blades

THERE HAVE BEEN MANY conscientious attempts to design beneficial operations for intractable asthma. A review of these endeavors demonstrates clearly the bewildering consequences of empirical surgical approach to disease of unknown etiology.

The history of various operations attempted for treatment of asthma are interesting. Probably the first to display an active interest in the autonomic nervous system's role in asthma was Kummell[1] in 1923. He advocated operations upon the vagus nerve but was fearful of the consequences of vagal interruption and decided to operate the sympathetic system instead; he performed a unilateral cervical sympathectomy. His associate, Braeucker,[2] influenced him in this decision since he had reported stimulation either of the sympathetic system or the parasympathetic system might result in bronchial constriction. There are many reasons, however, to question this conclusion. In 1924 Kappis,[3] also a German surgeon, divided the left vagus nerve just below the recurrent laryngeal nerve and was enthusiastic about his results. Leriche,[4] the famous pioneer of French surgery, performed stellate ganglionectomies for asthma and in addition destroyed portions of the sympathetic and parasympathetic nervous system.

Before various operations on the sympathetic and parasympathetic systems were employed, an operation was suggested as early as 1908 by Freund and Seidel which consisted of costal chondrectomy and transverse sternotomy. The application of this technique to human beings was based on operations used in horses to treat the "heaves," on the assumption that reduction of the barrel chest of both the horse and man might be beneficial. It is safe to predict that any clear-thinking horse would prefer the heaves, and there was little backlash from the unfortunate humans. The procedure has been abandoned. Thoracoplasty and phrenic nerve interruption were employed in the past. They are now obsolete.

For patients with emphysema associated with asthma, partial pleurectomy and pleural poudrage with talcum have been used in an effort to increase the blood supply to the lungs by collateral circulation between chest wall and the lung. The results have been indifferent.

The latest favorite operation for the treatment of emphysema and asthma has been the resection of the glomus caroticum (glomectomy). Theoretically, this procedure was devised to correct the autonomic imbalance of bronchial asthma. A theoretical consideration was that a low arterial tension stimulates respiration by way of the chemoreceptors. The procedure was popularized by Nakayma and adopted with enthusiasm by Overholt and other American surgeons.[5] This operation is based on the original theoretical consideration that if it has any merit at all, it is probably in patients with emphysema rather than asthma. It is, however, difficult to conceive a valid rationale for the operations, and as more experience has been gained, most surgeons have abandoned the procedure.

Early reports concerning the surgical treatment of asthma—and this applies to almost all of the operations attempted—have been characterized by enthusiasm completely dampened by follow-up studies that did not live up to the early expectations. Moreover, with the availability of modern drugs and the steroids, these heroic measures are not justifiable for the treatment of intractable asthma. There are currently only two absolute indications for operating upon patients with asthma or emphysema; both indications are complications of the disease and their surgical correction is without effect on the disease per se.

The first is spontaneous pneumothorax which may create a real emergency with the rupture of subpleural pneumatoceles. The other is a huge pneumatocele or multiple pneumatoceles which suppress or embarrass respiration and increase carbon dioxide. The results of surgery in this particular group are highly gratifying if the patients are properly selected.

Before drug therapy was effective, there had been modern revivals of operations of the autonomic nervous system for hopelessly intractable asthma. These include Carr and Chandler's reports[6] on dorsal sympathetic ganglionectomy, Phillips and Scott's operations[7] on the parasympathetic system and later Rienhoff's and Gay's contributions[8] concerning bilateral posterior pulmonary plexus resection.

It is easy to have initial enthusiasm for these procedures for what appears to be hopeless cases, and the studies by Blades and Beattie and associates[9,10] initially appeared to have significantly good results, not only from the subjective point of view but in the effect of pulmonary denervation on lung volume in asthmatic patients with emphysema. In this series the satisfactory results were comparable to those of sympathectomies for hypertension. There were, however, no reliable criteria for selection of patients for surgical treatment which correlated with final results.

PULMONARY BLEBS OR PNEUMATOCELES

Patients with asthma, particularly those with associated emphysema, are vulnerable to the development of subpleural blebs. In cases of obvious and fairly severe pulmonary emphysema, the pneumatoceles may be confined to a lobe or segment in which the emphysema is far advanced. These lesions may be bilateral or unilateral.

Subpleural blebs are commonly found at postmortem examinations and during intrathoracic operations for other diseases. They are asymptomatic. It is only after rupture of these lesions creating a pneumothorax or after tremendous enlargement of the pneumatocele which compresses the lung that clinical manifestations of the disease are evident.

In most instances, subpleural blebs are not visible on conventional roentgenogram because the superimposed lung tissue will mask their presence. If there is leakage from the pneumatocele and resulting pneumothorax, sometimes the lesion or lesions can be seen on a conventional roentgenogram.

The most common site for subpleural blebs is the apical portion of the lung, and another fairly common location is the superior division of one or both lower lobes. A serious complication in these circumstances is the development of a ball-value mechanism in the bronchial communications; air enters the bleb but cannot escape. If the bleb is intact it may be inflated to enormous proportions and create the situation of the so-called disappearing lung. The treatment of this serious condition is usually easy and gratifying, provided the correct diagnosis is made. After removal of the pneumatoceles the remaining normal or near normal lung tissue will usually inflate, and ventilatory function is dramatically improved.

Spontaneous Pneumothorax from Ruptured Subpleural Pneumatoceles

Spontaneous pneumothorax as a complication of asthma and emphysema is not common. It can occur, however, and in such a patient the urgency for treating it is much greater than in the nonallergic patient. The seriousness of this complication depends upon the amount and rate of air leakage creating the pneumothorax. If there is a large leak and a rapidly progressing tension pneumothorax, aspiration of the air from the pleura is an emergency life-saving procedure. In these patients it is usually advisable to insert an intercostal catheter attached to a water-seal drain to effect reexpansion of the lung. In some patients a small amount of air will leak into the pleural cavity and only a needle aspiration will be necessary. If there is any doubt, however, an intercostal catheter should be employed. In all circumstances these patients should be kept under very close observation because it is common for the air leak to stop and then recur as the lung reexpands.

This type of spontaneous pneumothorax and the rupture of subpleural blebs is common in the young and usually otherwise healthy patients. The first indication of the rupture of the bleb is chest pain followed by progressive shortness of breath, depending on the magnitude of the air leak. There is almost never a history of associated exertion; the attack may occur at night, waking the patient, or when the patient is at complete rest (Fig. 6-1) .

If there is a history of two or three attacks of pneumothorax or if decompression with the intercostal catheter does not cause reexpansion of the lung, an open operation must be performed. The blebs or pneumatoceles should be excised and any visible leak repaired. After this the lung is reex-

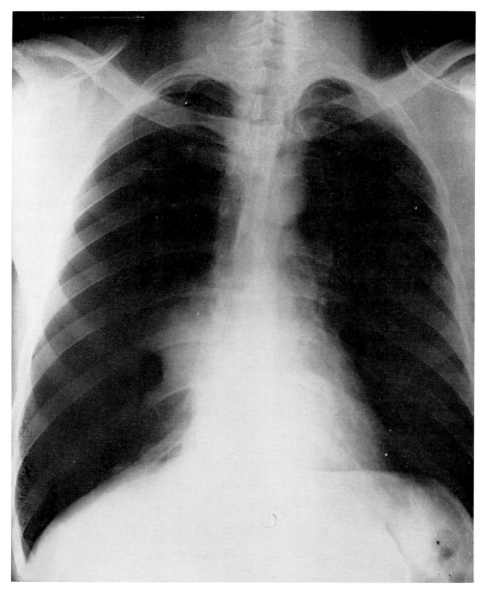

Figure 6-1. A roentgenogram of a patient with a progressive tension pneumothorax, following rupture of a subpleural pneumatocele. Decompression was accomplished by intercostal catheter attached to a water-seal drainage bottle.

panded and an intercostal catheter is placed in the pleura to insure removal of any trapped air or serum.

In a few cases one finds the greater part of the surface of the lung covered with small pneumatoceles. Excision of all these might result in damage to the lung surface; under these circumstances, the lung should be reexpanded and measures taken to create pleural adhesions. This may be done by rubbing the pleural surface with gauze or with a sterile talc poudrage. Some surgeons have advocated producing adhesions by pleurectomy; this greatly increases the magnitude of the operation and, in my opinion, is not necessary. Moreover, the complications in a pleurectomy can be quite serious, because of oozing from the raw surfaces created. In all cases, measures should be taken to create pleural adhesions.

Regardless of the type of operative intervention, extremely careful attention should be directed to the prompt reexpansion of the lung and maintaining it in an expanded position. This is done by use of an intercostal catheter or catheters attached to a water-seal drain. In most cases, the lung will be reexpanded in two or three days and the catheters can be removed.

Bilateral Pneumothorax from Subpleural Blebs

In 10 to 15 percent of the cases in which spontaneous pneumothorax has occurred, the condition may be bilateral. Ordinarily, it will occur on one side, and later the same condition may be present on the other side. Spontaneous pneumothorax bilaterally constitutes an extremely serious situation, and prompt bilateral decompression must be accomplished or the patient will suffocate. Recurrence of bilateral pneumothorax is a positive indication for open surgical treatment. Under the most common circumstances the operation should be staged so that the more severely affected side is operated upon first, and the second operation follows a month or so later. A few surgeons have advocated combining the two operations into one procedure for patients with bilateral subpleural blebs. In my opinion this is dangerous and should be employed only in patients with simultaneous bilateral pneumothorax. If the lungs do not reexpand with bilateral intercostal catheter water-seal drainage, a bilateral procedure is indicated. It is desired to make separate anterior thoracotomy incisions and avoid transecting the sternum, a technique which has been advocated by some. Except in the most unusual cases, when the disease is located in the posterior surface of the lower lobe, an anterior intercostal incision at the level of the second or third interspace is preferable to the posterior lateral approach.

Subpleural Blebs Associated with Localized Emphysema

The blebs or pneumatoceles associated with localized emphysema in asthmatic patients appear more frequently in middle-aged and older patients.

Whether the subpleural blebs common in the younger age group are a separate entity or an early stage of localized emphysema is not known.

Localized excision of the blebs, or segmental or wedge resections, are the operations of choice. Total lobectomy may occasionally be necessary but should be avoided if possible. It is obvious that conservation of lung tissue in this type of patient is of prime importance (Fig. 6-2, 6-3 and 6-4).

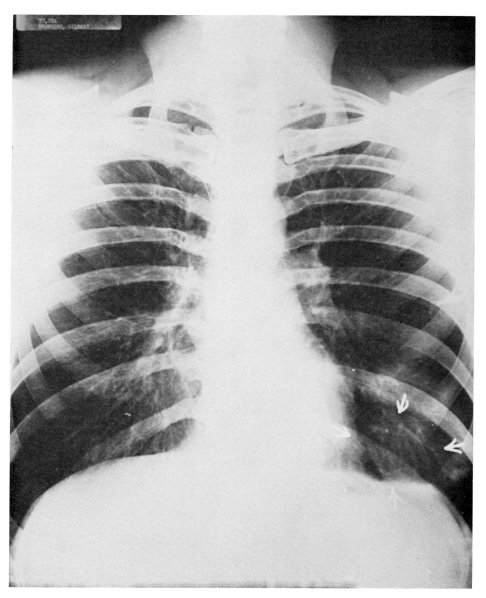

Figure 6-2. The roentgenographic appearance of bilateral pneumatoceles associated with asthma and emphysema.

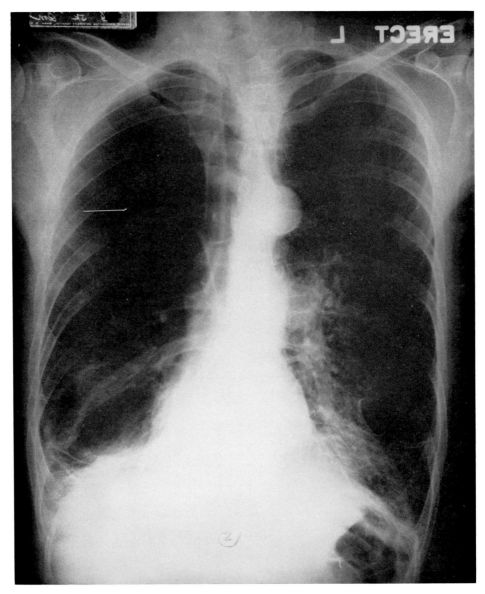

Figure 6-3. Large bilateral pneumatoceles are seen on the roentgenogram of a patient with long-standing asthma and emphysema.

Generalized Emphysema Complicated by Blebs

Respiratory embarrassment due to progressive inflation of a bleb is more common in patients with emphysema that is generalized rather than localized. The incidence of spontaneous pneumothorax in this group also is considerably less. In a recent series of twenty-four patients operated upon at

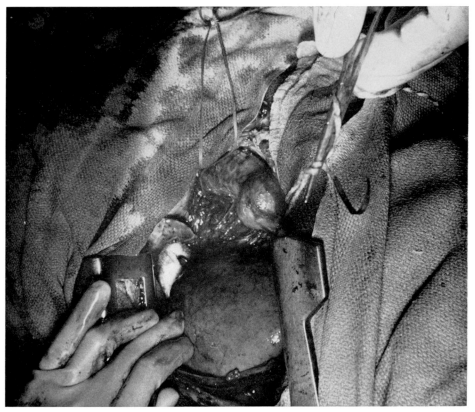

Figure 6-4. The appearance of localized emphysema and pneumatoceles exposed at operation. The lesions were excised by partial lobectomy.

the George Washington University Hospital, eleven had histories of recurrent attacks of pneumothorax, and in thirteen there was no history or evidence of previous pneumothorax.

It should be emphasized again that the operations and conditions just described constitute complications in the allergic patient, particularly those with emphysema.

Differentiation between primary emphysema and emphysema associated with long-standing asthma and allergic disease is of little practical importance as far as the surgical management is concerned. True allergic asthma usually appears at an early age with an abrupt and dramatic onset. Frequently, the initial attack is preceded by coughing which may wake the patient from sleep and is followed by a characteristic attack of asthma. True primary emphysema is usually manifested in older patients with a history of progressive shortness of breath. Physical signs and often the roentgen appearance of the lungs are similar to those of true asthma.

TRACHEOSTOMY

Tracheostomy is one of the oldest operations in surgery and is indeed one of the most important. With the availability of modern drugs the necessity for tracheostomy in the allergic patient is fortunately rarely required. But if there is any question about the wisdom of performing a tracheostomy, it should be done immediately. Tracheostomy produces a reduced respiratory dead space which is particularly helpful in the treatment of emphysematous patients, increases the ratio of tidal air to the dead space and reduces resistance to inspired air. In the patient with borderline pulmonary reserve, the operation may be lifesaving.

There has recently been an increasing tendency to employ endotracheal tubes in lieu of tracheostomy. In selected cases this may be a satisfactory substitute, but in allergic patients the presence of an indwelling endotracheal catheter can be very troublesome; in my opinion this technique should rarely be used. A tracheostomy should be performed.

The vertical incision is made in the trachea in the region of the second, third, or fourth cartilaginous rings. Some prefer to remove an eliptical-shaped piece of the wall of the trachea. A dilator may be employed to open the trachea and a suitable size tracheostomy tube is inserted. After the obstruction is removed, the inner cannula is inserted and locked in place, and immediate tracheal aspiration can be carried out with a soft rubber catheter.

It is of extreme importance that the wound be left open because any tight closure around the tracheostomy tube invites infection. A piece of dry gauze divided to encircle the tracheostomy tube can be placed under the tube and the tube secured with the tapes around the patient's neck. Usually, a size 5 to 7 tracheostomy tube is suitable for adults.

Emergency tracheostomies may be done anywhere with almost any available instruments. Such a situation usually means there has been a procrastinated decision to perform a tracheostomy. If performed under these circumstances, however, and assistants are not available, a vertical midline incision is advantageous. The fingers of the left hand can be employed to control bleeding and steady the trachea, which is exposed as quickly as possible. With the right hand the incision into the trachea then can be made. By twisting the knife blade or handle to enlarge the opening, the tracheostomy tube is inserted. Once the tracheostomy tube is in place, and an adequate airway established, all bleeding points should be identified and carefully ligated in order to obviate future complications.

Tracheostomies done under dire emergency conditions are much more dangerous and unsatisfactory. If there is a possibility that tracheostomy will be considered necessary sooner or later, the operation should be done in the operating room with standard aseptic techniques and with ample assistance.

Tracheostomy can be done under local analgesia. Except in an emergency situation of grave importance, a transverse incision in the neck is highly desirable. The incision is made about an inch or an inch and a half long and an inch above the sternal notch, depending somewhat on the build of the patient. The preference for this incision is based on cosmetic reasons especially important in women.

The skin edges and platysma muscles are retracted with small rake retractors and the deeper tissues divided in the midline. Veins running vertically can usually be avoided and those vessels in the transverse position ligated and divided. The strap muscles are retracted laterally exposing the trachea. In some instances, the isthmus of the thyroid gland may be encountered, and it may be retracted or, if necessary, divided between suture ligatures.

Postoperative Care of the Tracheostomy Patient

Postoperative management of the allergic patient who requires a tracheostomy is of utmost importance. Suction should be employed as often as necessary to keep the air clear. It must be remembered that careful observation is necessary at all times since the patient with a tracheostomy cannot call for help. It is desirable not to disturb the outer cannula for at least forty-eight hours. The inner cannula, however, should be removed and cleaned frequently. After a definite tract is established, the outer cannula can be safely removed and replaced if necessary.

If oxygen is necessary, it is better to give it by tracheostomy mask so that tracheal suction can be employed if necessary. If the secretions are minimal, then it is suspected that the tracheal bronchial tree is dry, a common condition if the humidity is low. Cold vapor can be given with a hydrojet or other apparatuses in an effort to prevent the inner cannula from being plugged with dried secretions.

When it is considered safe to remove the tracheostomy tube, two precautions should be taken: plugging of the inner cannula for increasing periods of time or, perhaps less desirable, substituting successively smaller cannulas. If the tracheostomy tube has been in place for a long time, the patient may become dependent upon it; sometime it becomes difficult to evaluate the true need for the tracheostomy. The common mistake is employing a tube which is too large. When this is occluded by plugging, it is impossible to breathe around the tube. The false impression that the tracheostomy is still essential may be erroneous.

Not infrequently, available tracheostomy tubes are too long. This is especially true in infants. If a tube of proper length is not immediately available, several gauze strips should be placed under the flanges to prevent the tip from entering a principal bronchus, usually the right. The tube that is too large and too long may sometimes erode the tracheal wall causing severe complications, usually in the form of hemorrhage.

Two safety measures are necessary. First, if possible, a roentgenogram of the chest should be obtained to ascertain the exact location of the tip of the tube; and second, any unusual movement of the tracheostomy tube with the pulse should be noted because a shorter tube is probably indicated.

BRONCHOSCOPY

Distal airway obstruction is usually the result of retained secretions. In adult allergic patients, these secretions may be thick and gelatinous, and if the patient is unable to cough material up effectively, an overwhelming pulmonary infection or atelectasis may result. If, therefore, usual measures to clean the distal airway, including catheter suction and bronchoscopy, have not been successful, tracheostomy is indicated immediately (Fig. 6-5).

One helpful technique used after bronchoscopy, in an effort to clean out the secretions by aspiration, is a tracheostomy done with a bronchoscope in place. During the procedure, oxygen can be given through the bronchoscope which is not withdrawn until the operating field is prepared for the immediate insertion of the tracheostomy tube. If the retained secretions are still a problem, bronchoscopy can be repeated easily through the tracheostomy tract as often as necessary. It disturbs the patient very little if done in this manner.

A cuffed tracheostomy tube is rarely necessary in allergic patients with retained secretions. Moreover, respirators such as the Berg respirator and the Engstrom unit should be avoided if possible in the allergic asthmatic patient because overexpansion of the lungs can be dangerous.

Anesthesia for Bronchoscopy and Tracheostomy

The subject of anesthesia in the allergic patient is reviewed in an admirable way in Chapter 8 and requires no comment, except that if local analgesia is to be employed, it must be employed with care, and if general anesthesia, an agent which is a known bronchodilator is desirable (this is particularly true in children).

BRONCHOGRAPHY

The roentgenographic study of the bronchial tree by the injection of radiopaque solutions is the only known method of determining the presence and the extent of bronchiectasis.

The necessity for bronchography in the allergic asthmatic patient is limited. Rarely, localized bronchiectasis appears to act as the trigger point for repeated asthmatic-like attacks in which case, a bronchogram is, of course,

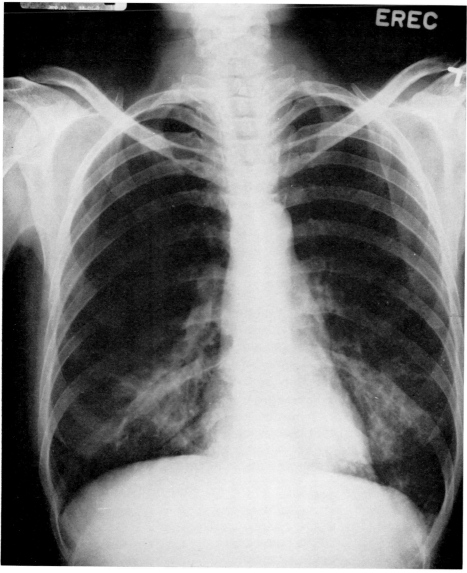

Figure 6-5. Bilateral patchy atelectasis in an allergic asthmatic patient.

necessary. Also, in the patient having bronchial stenosis bronchography aids in differentiation between pulmonary complications due to a pneumonia-like process and those due to atelectasis. If the lesion is acute, bronchograms may show what has been called pseudobronchiectasis, that is, some enlargement of the bronchi quite similar to cylindrical bronchiectasis but what is really the result of loss of muscle tone from the inflammatory changes in the lung.

It is usually necessary to perform bronchography under local anesthesia. The technique varies. If local analgesia is to be employed, the patient who has fasted for approximately twelve hours is prepared by giving him a small dose of barbiturate, and, if there is no contraindication, codeine. The barbiturate will furnish sedation and minimize a possible reaction to cocaine or other topical analgesia. In patients who have been coughing up large amounts of sputum, preliminary postural drainage is necessary.

It is exceedingly important that the procedure be described to the patient and each step be explained in detail, for his full cooperation is of utmost importance. Particularly should he be cautioned not to cough.

The larynx and pharynx are sprayed with the anesthetic agent (cocaine is used more than any other) while the patient holds his tongue forward and breathes deeply. A cotton swab dipped in the anesthetic agent and squeezed to remove any excess solution is used to anesthetize the piriform fossa. We prefer to employ for adults a No. 14 French catheter, which is passed through the nose previously sprayed with the anesthetic solution. The patient is instructed to take deep breaths, and the catheter is passed quickly between the vocal cords and into the trachea. This procedure can usually be done blindly without difficulty. In a few instances, it is necessary to visualize the larynx with a mirror, when the catheter is guided between the vocal cords with forceps.

Following introduction of the catheter, there is usually some coughing, and the voice is reduced to a whisper. The patient has already been warned that this will happen. The catheter is then taped in position and approximately 1 ml of cocaine solution or other agent is injected through it. Coughing spreads the solution throughout the tracheobronchial tree. Analgesic solutions should never exceed 5 to 7 ml.

In allergic patients caution must be employed in the installation of radiopaque solutions. Use of oily solutions, such as the Lipiodol®, should be completely avoided because of its retention and content. It is safe to state that the majority of the reactions to Lipiodol have been the result of the patient swallowing rather large amounts of the material, and free iodine is liberated when it reaches the stomach. Water-soluble solutions are much safer; propyliodine (Dionosil®) provides an excellent contrast material for about thirty minutes, after which it begins to fade, and in one to three days no trace of it can be seen in plain roentgenograms. It is necessary to shake the preparation well before use since it is a suspension and not a true solution. Warming is not necessary, as in the case of Lipiodol, and may indeed increase the chances of alveolar filling.

Bilateral bronchograms may be performed in one sitting, unless unilateral or selective bronchography of one lobe is desirable. Anterior, right and left oblique roentgenograms will make all lobes identifiable if the bronchogram is satisfactory.

When the contrast media is given through a catheter, small amounts of the material should be injected in intervals of one to two minutes; 15 to 25 ml contrast material is almost always adequate for bilateral bronchograms. The following positions are employed to delineate all lobes: (a) an upright position with the patient inclined posterolaterally, then laterally, then anterolaterally and finally anteromedially to fill the middle and lower lobes; (b) the Trendelenburg position, first supine, then in the lateral position in an effort to visualize the upper lobes. If the patient is in good physical condition, the knee-chest position can be used instead of the Trendelenburg position. The same sequence is used on each side.

Roentgenograms should be taken immediately after the injection of the contrast agent. The procedure may be repeated for visualization of the lobe or lobes not clearly delineated. After bronchography has been completed, the catheter is removed and the patient is urged to cough up all of the contrast material he possibly can. Unless the physical condition precludes it, postural drainage should be instituted intermittently for the remainder of the day.

Complications and reactions from bronchography could be caused by either the anesthetizing agent or the contrast media. The majority of cocaine reactions can be prevented by avoiding overdose. Regardless of the agents employed, arrangements must be made for the immediate administration of a quick-acting barbiturate and if syncope occurs, epinephrine should be available for injection. A rare, severe anaphylactic reaction, possibly with cardiac arrest, is to be considered, and the equipment and drugs necessary for treatment should be part of the standard equipment available, including the administration of oxygen.

Occasionally some hours after bronchography, symptoms and signs of allergic reaction with asthmatic attacks may be evident. This is much commoner when Lipiodol is used. These late reactions are rarely hazardous to life and are easily controlled.

When general anesthesia is necessary, usually in children, contrast media may be injected through the bronchoscope or by needle through the trachea and the required roentgenograms taken while the subject is still asleep. This in general is not a completely satisfactory method and it is much more difficult to obtain visualization of all of the lobes. Under these circumstances, if one lobe or lung is the target of suspicion, it is better to settle for visualization of this portion of the lung rather than attempt to accomplish complete delineation of both lungs.

BRONCHIAL STENOSIS

Bronchial stenosis is a rare complication in allergic patients. However, the patient with bronchial stenosis may wheeze and have an incorrect diagnosis of asthma. This mistake is made commonly in cases of bronchogenic carcinoma.

REFERENCES

1. Kummell, H., Sr.: Die operative Heilung des Asthma bronchiale. *Klin Wschr, 2:* 1825, 1923.
2. Braeucker, W.: Die experimentelle Erzenzang des bronchial Asthmas und seive operativ Beseitizing (anatomisch-chirurigische Studie). *Arch Klin Chir, 137:*463, 1925.
3. Kappis, M.: Die frage der operativen Behandlung des Asthma bronchiale. *Med Klin, 20:*1347, 1924.
4. Leriche, R., and Fontaine, R.: Resultate eloignes der traitement chirurgical de l'asthma bronchiqell. *Bull Soc Nat Chir, 54:*6600, 1928.
5. Overholt, R.H.: Resection of carotid body (cervical glomectomy) for asthma. *JAMA, 180:*809, 1962.
6. Carr, D.J., and Chandler, H.: High dorsal sympathetic ganglionectomy for intractable asthma. *J Thorac Surg, 17:*1, 1948.
7. Phillips, E.W., and Scott, W.J.M.: The surgical treatment of bronchial asthma. *Arch Surg, 19:*1425, 1929.
8. Rienhoff, W.F., Jr., and Gay, L.N.: Treatment of intractable bronchial asthma by bilateral resection of posterior pulmonary plexus. *Arch Surg, 37:*456, 1938.
9. Blades, B.B., Beattie, E.J., and Elias, W.S.: The surgical treatment of intractable asthma. *J Thorac Surg, 20:*584, 1950.
10. Blades, B.B.: The surgical treatment of asthma. *Postgrad Med, 6:*219, 1949.

Chapter 7

Surgery and the Allergic Pediatric Patient

William R. Richardson

THE ALLERGIC child is subject to all the types of trauma, disease and anomaly common to his age group. When and how the surgeon may contribute significantly to this child's medical care will be considered here. Attention is directed to special aspects of the management of diseases amenable to surgery as they appear in the allergic infant and child, with emphasis on those surgical problems that occur more frequently than in nonallergic children.

Although the surgeon's interest in the immune process in relation to organ transplantation has been revived, his role in therapy of any of the allergic states has been relatively minor. However, the importance of familiarity with certain conditions, such as the surgical complications of asthma, has increased because of their frequent presentation as emergencies.

ASTHMA

Specific Surgical Therapy

During the past fifty years numerous, sometimes bizarre, surgical procedures have been devised to alleviate some of the disabling manifestations of bronchial asthma.[1-4] Many of these seem to have evolved from a concept that there are several types of asthmatic patients, depending on the predominance of psychoneurogenic or infectious factors although supporting the importance of spasm, edema and secretions in raising airway resistance. Most of the operations have been abandoned, but disappointment with medical therapy and occasional benefits following surgical treatment continue to stimulate sporadic interest. Several procedures have been based on the attractive but unproved hypothesis that the person with asthma suffers from an inherent or acquired imbalance of the sympathetic nervous system.

Rienhoff and Gay (1938)[2] seemed encouraged by their early results in treatment of intractable bronchial asthma by bilateral resection of the posterior pulmonary plexus. They attributed "improved results" to application of the anatomic fact that the posterior aspect of the hilus is the only point at which all the efferent and afferent fibers composing the extrinsic nerve supply to the lungs can be interrupted. Miscall and Rovenstine (1942)[3] believed that their series of dorsal sympathectomies (interrupting pathways at stellate and second, third and fourth thoracic ganglia) demonstrated that "surgery is an important merited therapeutic procedure if it is restricted to patients with asthma of pulmonary origin who have been carefully selected after benefit from nerve blocking." The most recently publicized procedure has been unilateral carotid body excision (glomectomy), introduced by Nakayama in 1942.[4] Subsequent enthusiastic reports[5-7] have used questionably discriminating criteria in evaluating therapy, and several recent studies with controls, objective criteria, and even sham operations have demonstrated no benefits.[8-10] Bilateral glomectomy may be followed by a detrimental insensitivity to hypoxemia.[11] Reported experience with these specific procedures in children has been negligible, and a recent survey of thoracic surgeons by questionnaire revealed an overwhelming majority opposed to the theories and principles of glomectomy.[12]

Surgical Management of Complications

The role of the surgeon in current management of bronchial asthma is concerned less with manifestations than with complications, such as *extrapulmonary air leaks, respiratory failure, localized decreased aeration* (consolidation, atelectasis), *infection* (bronchiolitis, bronchiectasis), *bronchial obstruction,* and *deformities of the thoracic cage* (Fig. 7-1).

Subcutaneous and mediastinal emphysema and pneumothorax are not uncommon in asthmatic patients, although Bierman found only forty reported[13] from 1950 until his 1967 series of twenty episodes in sixteen children from Seattle, Portland and Spokane in a seven-year period.[14] Such

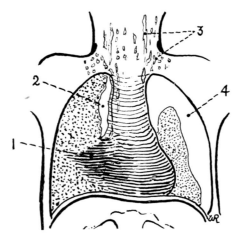

Figure 7-1. Diagram indicating some of the major "surgical" complications of asthma. (*1*) Right middle lobe syndrome (persistent atelectasis or consolidation), (*2*) pneumomediastinum, (*3*) emphysema of cervical fascial planes and subcutaneous tissue, and (*4*) pneumothorax. Other observed sites of extrapulmonary air are pericardial, paravertebral and peri-aortic, retroperitoneal and subserosal (pneumatosis intestinalis).

episodes have occurred more often in boys, the incidence peaks being in the preschool and adolescent years. Those in the younger group had milder asthmatic backgrounds but such associated pulmonary diseases as atelectasis, pneumonitis, bronchiectasis, or foreign body. The teen-agers usually had a more severe, chronic, recurrent intractable asthma without associated pulmonary disease. Of Bierman's thirteen patients under seven years, three had pneumomediastinum complicating the first asthmatic attack, and five others had had only mild to moderate asthma previously. The clinical picture is acute, featuring severe air hunger, anxiety and restlessness, hard brassy cough, and neck and chest pain. Common physical signs include absent cardiac dullness, subcutaneous crepitus and cyanosis. Bierman's follow-up revealed that about half his reported patients continued with moderate to severe asthma. Recurrent episodes of this complication are uncommon.

The postulated pathogenesis of pneumomediastinum involves a severe acute asthmatic episode with bronchospasm, edema and inspissated secretions leading to overdistention of alveoli and alveolar ducts. Contributing factors include limited elasticity of supporting structures, the shearing forces of respiratory stress, atelectasis secondary to mucous plugs, chronic steroid therapy, the forced inspirations of paroxysmal coughing and increasing use of intermittent positive pressure by mechanical means. Extension of mediastinal air into the neck is common in older children but rare in infants, who are more likely to develop pneumothorax. If the pressure gradient of the alveolar leak is not reduced, air may extend into one or both pleural cavities, subcutaneous tissues of the head or trunk or even to retroperitoneal or intraperitoneal locations. The sudden appearance of subcutaneous emphysema may at times be associated with significant relief of respiratory distress. As with all types of acute respiratory distress, a prompt roentgenogram is essential for accurate diagnosis and logical, optimal treatment. The lateral chest view is especially important to demonstrate significant mediastinal air between heart and sternum not always evident on the frontal film.

Management of pneumomediastinum basically depends on prompt, effective control of the asthma and cough that have led to alveolar overinflation. This alone usually permits absorption of mediastinal and subcutaneous air in five to seven days. Marked respiratory distress or an air-block syndrome in the absence of pneumothorax will usually benefit promptly from mediastinal aspiration by needle or catheter. X-ray-guided, oblique insertion of the needle or plastic catheter (Angiocath*) through the right costoxiphoid angle has proved simple and safe. If relief is not achieved or appropriate equipment is not available and there is increasing

---

*Available from Deseret Pharmaceutical Co., Sandy, Utah.

cyanosis and failing peripheral circulation in the presence of cervical sub-cutaneous emphysema, cervical mediastinotomy with or without tracheo-stomy may be dramatically effective.[15,16] A transverse suprasternal incision permits blunt midline exploration and decompression with a hemostat. Only if the situation also indicates an urgent need for reduction of respiratory dead space and facilitation of tracheobronchial aspiration should naso-tracheal intubation or tracheostomy be added. The recently developed micro-studies of blood gases and acid-base balance have provided a very useful means of monitoring the severity of the episode, the indications for, and the effectiveness of, therapy.[17]

In contrast to the alveolar and ductal leaks thought to be associated with mediastinal air, pneumothorax probably more often follows rupture of a pleural bleb by a sudden increase in transpulmonary pressure on a thinned pleural area due to a bronchial plug or by excessive pressure with mechanic-ally assisted respiratory ventilation. Absence of breath sounds, tympany, and even mediastinal shift are diagnostic signs in the very young patient. Respiratory distress from acute pneumothorax in the older asthmatic child is rarely as rapidly progressive as in the infant made vulnerable by such anatomic features as a flexible mediastinum. However, such a pneumo-thorax can seriously complicate management of the child already severely burdened by an acute severe epinephrine-resistant episode of asthma.

Sudden development of pneumothorax or pneumomediastinum can pro-duce a more severe clinical picture of distress than expected from the degree of actual ventilatory impairment. It is quite probable that complications from extrapulmonary air leak are more frequent than has been widely ap-preciated, because of the lack of diagnostic roentgenograms in many pa-tients. A chest roentgenogram upon admission to the hospital is important in all patients with status asthmaticus to rule out such complications since these must be treated prior to mechanical ventilation. The presence of either a volumetrically or manometrically significant pneumothorax, iso-lated or associated with mediastinal or subcutaneous air, is an indication for water-seal intercostal tube drainage to evacuate the air, restore pulmon-ary and pleural dynamics, and achieve complete reexpansion of the lung. Special care must be taken to diagnose and adequately treat the occasional bilateral pneumothorax.

Closed thoracostomy is used to establish intercostal tube water-seal drainage with or without suction (Fig. 7-2). The use of a well-equipped treatment room is preferred, and the operator and assistants should be masked, gowned and gloved. Local infiltration anesthesia with 0.5% pro-caine or lidocaine is used. If the setup is familiar and equipment pre-arranged, one assistant to restrain the child can suffice. Recent postero-

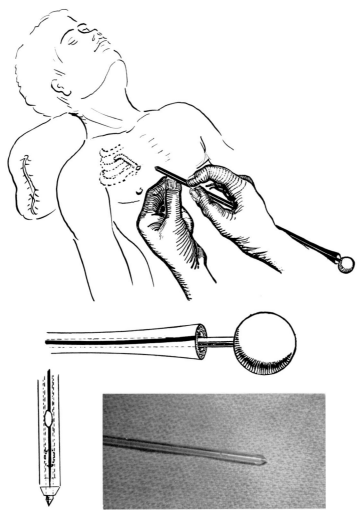

Figure 7-2. Simplified technique for institution of closed intercostal tube thoracostomy decompression using the recently available plastic trocar catheter with sentinel eye and radiopaque sentinel line.* (*Top*) Oblique insertion route to lessen chance of air leak about the tube can be developed with hemostat, but sharp inside trocar point makes a small, efficient pleural puncture just above the chosen rib, preferably second or third anterior, with child in Fowler's position. (*Bottom*) Close-up of catheter features. Sizes available are 10 Fr. to 32 Fr. (*Argyle trocar catheter, Aloe Medical, St. Louis, Mo.)

anterior and lateral chest x-ray films should be available to guide a more precise placement. The second interspace anteriorly in the midclavicular line is preferred, with the patient appropriately restrained in Fowler's position. Skin and subcutaneous tissue are anesthetized with an intradermal needle which is followed with a short bevel 22-gauge needle down to the

rib and then redirected upward through the next intercostal space and be-
neath the endothoracic fascia, for injection of additional anesthetic solution.

For years we used a straight rubber urethral catheter provided with one
or two extra holes and introduced via trocar, or more often by simple hemo-
static "punch" insertion.[18,19] Recently there has become available an excellent disposable plastic catheter trocar set which has the advantage of
utilizing small puncture wounds of the skin and pleura, and a radiopaque
line on a transparent, minimally reactive tube. Its sterility is more certain
and it probably costs less (Fig. 2b). Fixation of the catheter to skin by
suture or adhesive assumes greater importance in children since it avoids
skin-level angulation or displacement by sudden tug. The catheter is
promptly connected to a previously assembled three-bottle or appropriate
suction drainage system, usually with 10 to 12 cm of water at negative
pressure. Water-seal drainage without suction is relatively ineffective in
infants because of their limited tidal volume. If appropriate tube equipment is not available for the occasional sudden, severe tension pneumothorax, the widely available large Angiocath or the Intracath* plastic
catheter is preferred to the plain needle and may easily be adapted for
water-seal drainage with a modified Baxter Vacoliter flask and intravenous
tubing.[20] An immediate postintubation chest x-ray film is important to show
position of the tube and the effectiveness of reexpansion. Continued leak
of a small bronchopleural fistula has not been a problem; tubes are usually
removed forty-eight hours after an air leak ceases if the lung has remained
expanded and asthmatic control has improved.

Respiratory Failure

Pediatric intensive care units, originally intended for the care of complex postoperative or neonatal patients, have evolved into areas where care
is also provided for severely injured and acutely ill infants and children.
Now a patient may be categorized by the degree as well as the type of his
illness, and a newly labeled specialist—"the intensivist"—has been named to
head the team required for proper application of physiological knowledge,
complicated equipment and skilled personnel.[21] Inhalation therapy and
respiratory care have increasingly become basic functions of such units, en-
compassing aspects of anesthesiology, internal medicine and pediatrics,
physiotherapy, general and thoracic surgery and specialties such as allergy
and otolaryngology. The surgical specialist's role is that of secondary or
consulting team member in the care of respiratory failure in the asthmatic
child.

The diagnosis of respiratory failure demands that the physician take

---

*Available from Deseret Pharmaceutical Co., Sandy, Utah.

immediate steps to augment alveolar ventilation. Mechanical ventilation
with an artificial airway has recently been advocated by several centers as
the most effective method for managing acute respiratory failure in both
status asthmaticus and acute bronchiolitis.[22-24]

Establishing Artificial Airways

Three procedures are available, nasotracheal intubation, standard tracheostomy and cricothyroidotomy. The last hardly deserves discussion here
since appropriate pediatric equipment is rarely available and its only real
benefit is sharply limited to extreme upper airway obstruction, a rarity in
asthmatics of any age. Both intubation and tracheostomy can produce several
desirable results: (a) a decrease in extrapulmonary air leak by reducing
tracheobronchial pressures, (b) reduction of dyspnea by reducing dead
space, and (c) facilitation of aspiration of secretions. Tracheostomy can
also provide mediastinal decompression via the cervical wound.

The following disadvantages of tracheostomy have contributed to the
current surge in popularity of prolonged tracheal intubation: (a) as a true
operation it requires local or general anesthesia; (b) its hazards in a sick
child may contribute to indecision and delay; (c) reported operative complications are significant, including bleeding, airway obstruction, pneumothorax and pneumomediastinum; (d) maintenance, especially in infants, is
truly hazardous, including mechanical problems with the tube and ventilator, accidental extubation, and mediastinal or tracheobronchial infection;
(e) extubation, especially in infants, can be difficult and late scarring may
occur. Repeated espisodes requiring mechanical ventilation by tracheostomy
could result in tracheal stenosis.

Extensive trial of prolonged nasotracheal intubation[25] in the last ten
years has clarified some of its advantages and disadvantages. Because the
procedure is easy and brief, it encourages a prompt decision to establish an
artificial airway and this contributes to minimal immediate risk. Complications during the period of intubation and following extubation, however,
have proved significant in children with upper airway disease. These include subglottic stenosis and membranes, hoarseness, tracheal stenosis and
pressure necrosis of the ala nasi. This incidence of complications contrasts
with the low one reported by Striker *et al.*[25] in patients intubated for
respiratory failure from causes other than primary obstructive airway disease (3 complications in 98 children). They found nasotracheal intubation especially valuable in conjunction with mechanical ventilation for the
treatment of respiratory failure caused by recurrent episodes of status
asthmaticus.[22] Continued nasotracheal intubation is more likely to require
sedation of the patient; obstruction of the tube by plugs, and increased

ventilatory resistance are more likely because of the greater length and smaller diameter of the tracheal tube.

In the current discussions of the relative merits of these procedures in the severely ill child with acute asthma, one should not lose sight of the really important question, When should the definite step toward an improved airway be taken? True indications consist of the signs that determine when this step should be taken. Increasing stridor, hoarseness, retractions and labored respiration are suggestive but not definite. More compelling signs are those reflecting failure of compensation and the appearance of true hypoxia or CO_2 excess. These include increasing restlessness, progressive significant pulse elevation, dulled responses with refusal to eat or drink and cyanosis in 40% oxygen. Blood-gas alterations are even more definitive and have provided an approach toward truly objective indications for intubation or tracheostomy, e.g. pH falling below 7.25, pO_2 below 50 mm Hg, or pCO_2 exceeding 65 mm Hg.[26] It cannot be overly emphasized that respiratory complications of severe asthmatic attacks such as pneumothorax or pneumomediastinum must be recognized and treated prior to mechanical ventilation by whatever airway access.

Our own view of mechanical ventilation has been conservative; we consider nasotracheal intubation with a polyvinyl chloride plastic* tube the method of choice for establishing an artificial airway for three or four days in infants and children with postoperative respiratory distress who are free of upper airway disease. In an emergency, an orotracheal tube is used initially and later replaced with a nasotracheal tube if the indication persists. Nasotracheal intubation is preferred to orotracheal because it avoids obstruction by biting, minimizes movement during swallowing, facilitates oral feeding and mouth care, reduces patient discomfort and is easier to secure and stabilize, with or without ventilator. In a child older than two years and sick with upper airway disease, three or four days' trial may be enough to indicate the desirability of tracheostomy. In a small infant an experienced team observing meticulous technique and intensive nursing care may advantageously continue intubation for two to three weeks or more. Use of a tube of proper length and diameter is exceedingly important. The external diameter of the tube is usually regarded as the one next smaller than that which traverses the glottis without resistance, thus allowing some leak of air and secretions. For each patient, the predicted size, together with the next smaller and the next larger, are always prepared, so some of the seemingly crude rules-of-thumb may be practical. Although the table-reference method is safest (Table 7-1), nostril size is a useful guide to the size of the glottis in infants,[26] and the rule, "tube size (French)

---

*Portex tube, S. Smith & Sons, Ltd., Jamaica, Long Island, N. Y.

TABLE 7-1

GUIDE TO CHOICE OF PEDIATRIC TUBE AND SCOPE SIZES

AGE	BRONCHOSCOPES		TRACHEOSTOMY TUBES			ENDOTRACHEAL TUBES		
			Metal		Plastic			
	Internal Diameter (mm)	Length (cm)	External Diameter	Holinger	Silastic	Internal Diameter (mm)	Fr. Size	Length (cm)
<4 mo	3.0	20	4.5	00	1.3	3.0–4.0	12–16	10–11
4–8 mo	3.5	25	5.0	0–1	3	4.5	16–18	11–12
8–18 mo	4.0	30	5.0–5.5	0–2	3	5.0	20	12–14
18 mo–5 yr	4.5	35	5.5–6.0	1–2	4.5	5.6	20–24	14–18
5–12 yr	5.0	40	6–8	2–4	6	6.7	24–30	18–22

equals in years plus 20," is fairly reliable in the first two decades.[27] Another "pearl" states that one can expect the appropriate tube to be as broad as the patient's little finger and as long as one and a half times the distance between the nose tip and earlobe.[27] The distance of insertion (nares to midtrachea) is estimated in advance as 21 percent of the infant's crown-to-heel length, but we must check by auscultation and roentgenography. For intubation by the nasal route, endotracheal tubes should be about 20 percent longer than those used for oral intubation.

We have avoided tracheostomy in most infants primarily because of the increased hazard of maintenance care and the great difficulty encountered during decannulation. However, recent improvements in technique and equipment have solved many problems.[27-29] Modifications of the tracheostomy technique for adults that we have found successful in infants deserve review here (Fig. 7-3).

To be consistently adequate, tracheostomy in an infant requires the airway control and tracheal identification provided by preliminary endotracheal intubation and is facilitated by general anesthesia administered in an operating room. The skin incision can be transverse or vertical, but thereafter all dissection should remain vertical midline with the thyroid

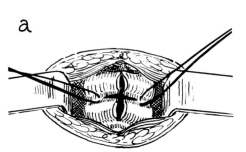
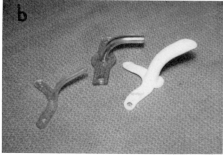

Figure 7-3. (*a*) Tracheostomy wound just before insertion of cannula. Skin incision is transverse, but all other dissection is vertical midline. The thyroid isthmus is retracted upward or divided. Lateral silk sutures* for operative stabilization will be left in place 5 to 6 days to permit easy access to the trachea in case of inadvertent extubation. Excision of a segment of tracheal wall as a window for cannulation is avoided. (*000 silk looped around one or two tracheal rings is more reliable than that illustrated.) (*b*) Examples of improved plastic tracheostomy tubes which seem to have specific advantages over standard silver tubes in infants and children. The white Silastic tubes of modified Aberdeen design have several desirable material qualities (radiopacity, "slick" adherent-resisting surface, autoclavability, nonirritabiliy, elasticity, softness and pliability). The latter two seem slightly excessive since they require too much wall thickness to maintain internal diameter. Pictured is a size No. 6 (I.D. 7.0 mm x O.D. 10.0 mm) with a.p. depth 1 $^{13}/_{16}$" and face plate 1 $^{1}/_{4}$" x 2$^{3}/_{16}$". The Portex clear plastic tubes shown are two of the smallest available, 4.5 and 7.0 mm O.D., respectively. Length can be trimmed and lumen partially unroofed if need be.

isthmus usually retracted upward. The endotracheal tube stabilizes the tiny trachea in the midline and can protect it against back wall injury during tracheal incision. Silk stay sutures are placed in the tracheal fascia or around the third and fourth tracheal rings on both sides of the midline. Insertion of the cannula through a vertical midline incision spanning two or three tracheal rings avoids some of the instability or collapsibility following decannulation which in infants often follows excision of a segment of cartilage. Insertion trauma and ring distortion are minimized by extending the incision through a third or even fourth ring or by making the opening cruciate with a short transverse slit in the membrane between the divided third and fourth rings. The incision is closed loosely if at all, and the silk tracheal traction sutures are left in for five to six days to facilitate the otherwise hectic if not hazardous replacement of an accidentally dislodged tracheostomy tube.

While the silver cannula with removable inner liner and obturator is still the most widely used, the well-designed plastic or Silastic® tubes now available* are more satisfactory because they are less irritating to the trachea, are pliable, may be trimmed to conform in contour and length, and simplify nursing care by requiring no inner cannula and by being shaped to facilitate or obviate wound dressings. It is rarely possible to use a cuffed tube in children under three years of age because of the small tracheal lumen (Table 7-1).

Thoracic Deformities and Asthma

Thoracic deformities for which asthma has been blamed are depression and protrusion deformities in children and a barrel-shaped chest in adolescents and adults. It is widely accepted that the latter may result from severe emphysema that inevitably accompanies chronic uncontrolled asthma, but it rarely develops prior to the adult years and has little significance for the surgeon except as a physical sign. Many developing thoracic cage distortions in the young should be reversible if maximal bronchodilation can be maintained, but the persistently elevated functional residual capacity benefits from special exercises to combat the reflex hypertonicity of the inspiratory muscles.[30,31] There is little to suggest that either pectus excavatum (funnel chest) or pectus carinatum (pigeon breast) should be considered a complication of obstructive lesions in the upper or lower respiratory tract rather than a congenital anomaly.[32] Derbes et al.[33] believed that later in childhood when chest cage rigidity has increased, a pigeon breast could result from

*Plastic from J. G. Franklin and Sons, Ltd., London, England, and Portland Plastics, Ltd., Hyde, Kent, England; Silastic from Medical Products Division, Dow Corning, Midland, Michigan.

the continued forced inspirations of uncontrolled asthma. This deformity features a lateral depression of the costal cartilages which produces the major physiologic problem, a substantially reduced intrathoracic volume. Currarino and Silverman[34] have demonstrated a premature fusion of the sternal segments in patients with pigeon breast. These sternal deformities are commonly observed at birth but are occasionally encountered in asthmatic patients.[31,35] The substernal ligament originally stressed by Brown[36] is rarely found at operation, and procedures limited to dividing this alleged diaphragmatic attachment have not proved successful. In the infant, where the chest wall is still soft and flexible, paradoxical inspiratory depression of the sternum is a conspicuous feature of deformity.

The characteristic "cough fractures" occasionally occurring in patients with bronchial asthma are found in the diagonal line of the interdigitating and opposing serratus anterior and external oblique muscles.[37] I have found none described in children.

PULMONARY COMPLICATIONS

Bronchiolitis

Relatively recent clarification of the relation and differences between acute viral bronchiolitis and bronchial asthma[38] has indicated that a fourth to a third of infants with an acute attack of bronchiolitis exhibit a major respiratory allergy and a high incidence of bronchial asthma or nasal allergy. With the exception of the rare patient with severe respiratory distress, most infants with acute bronchiolitis with or without allergic significance can be managed "conservatively." Bronchiolitis, in its purest form, is a disease of the periphery of the lung and should not be particularly helped by tracheotomy. If exhaustion or periods of apnea necessitate respiratory assistance, nasotracheal intubation is preferable for the first few days. If excessive tenacious secretions persist from pneumonia, bronchitis, or tracheitis, then tracheostomy may become mandatory. Perhaps it is pertinent that Wittenborg and associates[39] have reported expiratory collapse of the trachea in cineroentgenography of both acute bronchiolitis and asthma. The removal of inspissated mucus by bronchoscopy is less likely to be successful in bronchiolitis than in asthma, because of the former's more distal airway involvement. In severe cases of bronchiolitis, as in bronchial asthma, extrapulmonary air leaks may develop suddenly and complicate the management.

Consolidation, Pneumonia, Atelectasis and Collapse

The increasing frequency with which the chest roentgenograms are made in the allergic asthmatic child aids in excluding conditions simulating asthma and is providing a more accurate picture of the incidence of pulmonary complications with infiltrates. Several published series of pediatric cases[40-43] indicate how frequently an allergic problem underlies recurrent pneumonia and/or atelectasis in children. More intensive workup of allergic patients with pulmonary infiltrates has begun to clarify some of the "pulmonary hypersensitivity states with eosinophilia."[44] Bronchograms have proved sufficiently reliable and valuable diagnostic aids to warrant their early and repeated use to delineate developing anatomic changes. Bronchoscopy has not been as useful as a diagnostic test but is mandatory to rule out foreign body, mucous plugs, and endobronchial disease.

Several cases of acute collapse of the middle lobe have been encountered in allergic children.[42] Most such patients will respond to appropriate vigorous treatment of the underlying condition with antibiotics, expectorants, bronchodilators and mucolytic agents. These are supplemented by intermittent positive pressure breathing by respirator plus early, repeated tracheal aspiration with a maximal caliber suction catheter during direct laryngoscopy (Fig. 7-4). Bronchoscopy may be necessary to rule out foreign body or to achieve expansion of the lobe by relieving bronchial obstruction with or without bronchial lavage. Fortunately, few foreign bodies (3%) invade the right middle lobe, the predominant site of allergy-related atelectasis.

One Scandinavian author[43] found that bronchial asthma was a more common cause of recurrent pneumonia-atelectasis than mucoviscidosis prior

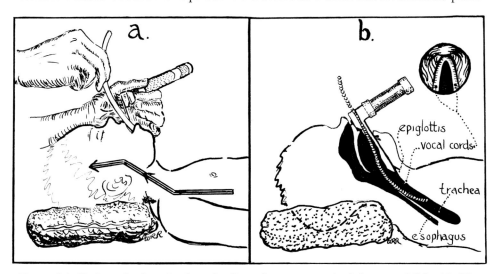

Figure 7-4. Endotracheal aspiration via direct laryngoscopy in infant or child. (*a*) The "amended" position of the patient (neck flexed, head extended on small firm pillow) is much preferred to the hyperextended or "classical" position. (*b*) Diagrammatic section of aspiration catheter inserted under direct vision. A slightly curved catheter segment attached to the tip of a Samson aspirator has proved especially easy to direct.

to 1966; it occurred in 14 of the 125 asthmatic children studied. It was twice as common in girls and had a strong predilection for the right middle lobe, as Dees and Spock[40] had emphasized in their report on thirty children with the right middle lobe syndrome seen at Duke University Medical Center in a ten-year period. Brock *et al.*[45] first stressed the surgical import-ance of disease in the right middle lobe (RML) in 1937 and in 1946[46] de-scribed the anatomic features that predispose this lobe to easy obstruction. The bronchus to the RML is unique for its length, small caliber, acute angle of branching from the right main stem bronchus, and its proximity to an encircling group of three or more lymph nodes which drain the right middle and lower lobes. Graham *et al.*[47] in 1948 coined the term *right middle lobe syndrome.*

Early reports of this syndrome included few children and dealt mainly with atelectasis and bronchiectasis of the RML associated with tuberculosis or hyperplastic lymphadenitis. Today many reported cases of the syndrome in children involve allergy, and significant constrictive lymphadenopathy is rarely associated with tuberculosis (Table 7-2). Probably a combination of factors are involved: allergic, infectious, or viscid secretions, bronchial mucosal edema, and lymph node compression in an anatomically vulner-able structure. Dees and Spock[40] believe that RML disease in children is much commoner than stated in the literature and that its diagnosis is de-layed or missed by lack of specificity of symptoms and the absence of signs distinguishing it from other allergic and infectious chest diseases. In adults, cough, fever, purulent sputum and hemoptysis are primary features. Chil-dren have cough, wheezing, asthma, dyspnea and recurrent pneumonia or bronchitis localized in the RML. For example, refractory respiratory symp-toms and x-ray evidence of decreased RML volume in a female "asthmatic"

TABLE 7-2

ETIOLOGY OF ATELECTASIS IN CHILDHOOD

Pneumonia
Operation
Infectious lymphadenopathy
Cardiovascular compression syndromes
    Vascular rings
    Intracardiac defects with increased pulmonary blood flow
Allergy
    Asthma
    Heiner's syndrome
Metabolic diseases
    Mucoviscidosis
    Hypogammaglobulinemia
    Kartagener's syndrome
    Immunosuppressive therapy
Foreign body
Tumor

Note: Table is modified from Tarney *et al.*[41]

child should furnish a clue to early detection of the syndrome. Symptoms have usually begun at one to two years of age, but the pathologic condition has not often been appreciated before age six years. The importance of establishing an underlying diagnosis is obvious.

In seems that cystic fibrosis is also being encountered more frequently as a cause of chronic atelectasis and infection, and an increase in palliative resection of severely involved areas of the localized disease in carefully selected patients has been noted.[48] The severity of the localized bronchiectasis, abscesses and peribronchial lymphadenopathy encountered in resection, plus the occasional success achieved by repeated bronchoscopy,[40,41] suggest the value of more aggressive efforts to achieve early reexpansion of the lung. Bronchoscopic aspiration and lavage with mucolytic agent have failed to clear areas of long standing atelectasis. The right middle or right upper lobes, or both, are most commonly involved. In Kulczycki and associates,[49] follow-up of 266 live patients, 44 had proved respiratory allergy and most of these, 36, a positive family history of allergy.

Therapy for atelectasis must seek prompt reexpansion of the lung and prevent progression of the disease. It is interesting to note that in allergic children without lung resection,[40] mean duration of respiratory symptoms

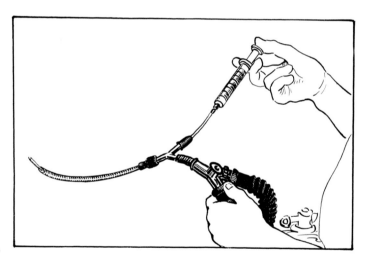

Figure 7-5. Apparatus for closed-system bronchography under general anesthesia is recommended by Smith.[26] Y-tube is interposed between endotracheal tube and anesthesia apparatus. Plastic catheter for instillation of opaque material is inserted through a snug perforated rubber nipple on the Y-tube and down the endotracheal tube. This permits selective positioning of the catheter, slow injection and serial films with better radiologic technique under such control. An esophageal stethoscope allows the anesthesiologist important and constant monitoring. Ventilatory exchange must be watched closely during injection of the dye and as much dye as possible aspirated through the endotracheal tube after the procedure.

before diagnosis of RML disease in those who improved was three and a half years compared to seven years in those who did not improve. Symptomatic therapy is the same as for collapse of the middle lobe. Repeated efforts at aspiration may be necessary to achieve any lasting expansion, and recurrent attacks are reported[41] to have the usual good immediate but transient response.

Measures to correct specific abnormalities are of obvious importance; cardiovascular surgical procedures, desensitization or removal of allergens, and appropriate antibiotic treatment have often made resection unnecessary. Dramatic relief has been observed in cases of Heiner's syndrome when milk has been withdrawn.[41,50]

Extirpative surgery should be reserved for those lung segments failing to show improved function after several months of therapy and having evidence of irreversible anatomic damage or resistance to repeated vigorous efforts at reexpansion. In about half of the reported patients with allergy who underwent lobectomy,[40,41] excised lobes have shown the more extensive pathologic changes of bronchiestasis in addition to the contracted volume predicted by x-ray and physical findings. Thus far these children have done well; symptoms of mild asthma and hay fever have continued but have remained controllable with medication previously ineffective. When intensive medical management fails to obviate the necessity for surgery, it should be realized that there is currently no evidence that compensatory regeneration or replacement of resected lung tissue occurs except possibly in small infants, and that if obstructive disease is present in the remaining lung tissue, dilation with emphysematous changes of remaining lung may be aggravated by excision.

Lung biopsy by limited thoractomy may occasionally be helpful in a child with diffuse parenchymal disease seen on roentgenogram.[51,52] Gibbon's list[53] of diffuse pulmonary lesions included fifteen conditions "due to allergy or of uncertain etiology," and only about a half to two-thirds of them can be diagnosed by tests other than histologic examination of lung tissue. Percutaneous needle biopsy of the lung has also proved its value and in appropriate cases should be tried before an open operation.[54] A recent report on needle biopsies showed a return of diagnostically useful tissue (77%) comparable to that reported for open-chest lung biopsy but accomplished with significantly less morbidity and hospitalization. Pertinent pediatric experience is still limited although Gellis advocated aspiration lung puncture some seventeen years ago.[55]

A qualified surgeon should perform open lung biopsy in appropriate children when risk is outweighed by the therapeutic advantage possible only with a tissue diagnosis (Fig. 7-6). The biopsy site varies with the

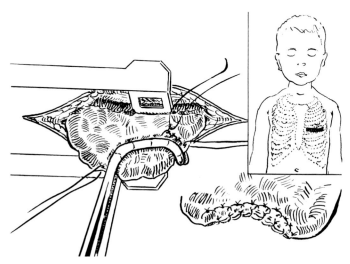

Figure 7-6. Open lung biopsy is occasionally useful in the diagnosis of diffuse pulmonary disease. Insert indicates our preference for a short anterior incision through an appropriate second to fifth intercostal space, use of a partial exclusion minimal-trauma vascular clamp to maintain inflation until fixation, overlapping interrupted nonabsorbable mattress sutures placed behind the clamp prior to excision, and reinforcement with a row of fine interrupted sutures to seal the edge of the wound.

suspected pathologic condition but is usually the inferior margin of the upper lobe. A 5 to 6 cm incision is made in the submammary intercostal space (4th to 5th), and a tip or wedge of lung is excised. The lung wedge defect is sutured meticulously to be airtight, the wound closed and tube drainage of the pleural space provided for forty-eight hours. The tip or wedge specimen should be resected with occlusive clamps in the inflated state and part of it immediately fixed. Complications are rare but pneumothorax, subcutaneous emphysema and bleeding have been reported. A chest roentgenogram is advisable a few hours after operation.

Clinical improvement following bronchoscopic aspiration of secretions with or without bronchopulmonary lavage[56,57] has been favorably reported in adults with status asthmaticus, bronchitis and emphysema, and pulmonary alveolar proteinosis. However, the limited volume and mucoid quality of tracheobronchial secretions in the young asthmatic child differ considerably from the copious, thick, mucopurulent secretions of the asthmatic adolescent or adult. I concur with Downes *et al.*[22] who have found bronchoscopy seldom of significant benefit to children with acute status asthmaticus. Dees[58] commented that bronchoscopy in an acutely ill child is often debated, rarely done with any enthusiasm by the endoscopist, and yet at times seems to be a decisive factor in clearing the airway and contributing to improvement. Fortunately, progress in medical management promises to shrink the need for bronchoscopy. Of particular importance have been management of hypoxemia and acidosis, mechanical ventilators and nasotracheal intubation for respiratory failure, and refined inhalation therapy techniques.

Bronchoscopy is even less likely to be successful in removing inspissated secretions in bronchiolitis than in asthma because of the more distal involvement in bronchiolitis. Richards and Siegel[23] believe that the procedure lacks the elements of ventilation and is an inadequate means of suctioning the distal radicles of the bronchial tree in status asthmaticus. Age factors like the increased risk of local complications plus the advisability of using general anesthesia in conjunction with bronchoscopy, have undoubtedly helped relegate this procedure to the background. Even Chevalier Jackson stated,[59] "It has generally seemed best not to do bronchoscopy during the asthmatic attack." Gibbon[53] noted that bronchoscopic suction could be helpful in certain patients with intractable asthma and severe emphysema who are unable to cough up secretions because they cannot generate enough tussive squeeze or they have a redundancy of bronchial mucosa. Fortunately, these stages of the disease are rarely reached in the pediatric patient. Speer and Todd[60] believe bronchoscopy is seldom necessary in children and that indications include (a) failure to respond to other procedures, (b) persistence of unilateral pulmonary symptoms, and (c) presence of a large area of atelectasis.

OTHER INDICATIONS FOR SURGICAL CONSULTATIONS

Miscellaneous indications for surgical consultation or treatment of allergic children occur with increased frequency. They include the differential diagnosis and treatment of wheezing and stridor, complications of prolonged steroid therapy such as bleeding or perforated peptic ulcers, edematous airway obstruction due to insect bite or allergy, and acute or recurrent abdominal pain.

Abdominal pain and gastrointestinal bleeding are always mentioned in reviews of allergic manifestations in infants and children, but both are rarely of surgical significance even if supported by valid evidence rather than speculation. "Colic" in infants is often attributed to milk allergy, and since elimination of milk from the diet has been curative it certainly deserves testing. Gryboski[61] recently reported that two infants had ileocolic intussusceptions reduced surgically during a period of trial milk feeding. Rubin[62] reported a rather well-defined clinical pain syndrome observed in a three-year period in six 3- to 5-week-old "colicky" infants with red rectal bleeding. Two had visible, large gastric peristaltic waves. Stool frequency and mucus were increased. A moderately severe anemia from bleeding developed in one. Response to omission of cow's milk was dramatic, with relief of pain and bleeding and return to normal in twenty-four to forty-eight hours.

Allergic or eosinophilic peritonitis with ascites in children has been reported[63] as well as acute pancreatitis found at autopsy. The latter is more likely the result of steroid therapy.[23] Laparotomy and liver biopsy have at times helped to diagnose allergic granulomatosis.[61,64] Some of the gastrointestinal manifestations of allergy may suggest a malabsorption syndrome that might simulate those manifestations occasionally encountered with surgical conditions such as malrotation of the colon, intestinal stenosis, chronic partial small bowel obstruction, or the adjustment phase of intestinal dysfunction that follows portal vein thrombosis in infants.[65] Andresen has for many years supported a major role for milk allergy in ulcerative colitis.[66] It behooves the pediatrician, allergist and surgeon to keep these clinical possibilities in mind so that no serious surgical problem will be obscured by attempts at desensitization and no allergy will be treated unsuccessfully or disastrously with operation.

In general, it should be anticipated that children with obstructive lung disease, even if well managed before, during and after surgery, will still have increased postoperative respiratory problems. In the infant age group, there is frequently respiratory depression because of limited respiratory reserve, small airways and the predominant diaphragmatic component of their breathing mechanism. The need to limit the diaphragmatic splinting effects of abdominal distention, tight dressings, or painful wounds is obvious. When a pain-relieving drug must be used, it is administered intravenously in approximately one-tenth of the intramuscular dose. In the older child, every effort is made preoperatively to achieve maximal pulmonary function. Preoperative explanation and demonstration of respiratory care will help to eliminate apprehension and promote cooperation. Larger children can be taught breathing exercises.

Retained secretions are mobilized with ultrasonically nebulized water, two-thirds normal saline and bronchial dilators. Intermittent positive pressure breathing may help introducing aerosols as well as chest physical therapy. The latter is extremely important in postural drainage. In pediatric intensive care, chest physical therapy can be administered by the therapists under the supervision of a nurse respiratory therapist. The aim should be removal of secretions by postural drainage and by manual assistance to coughing, through vibration and percussion followed by tracheobronchial suction.

Pediatric anesthesiologists have stressed the importance of following the basic rules of respiratory therapy during the period of anesthetic management, especially when operation must be performed in the presence of acute or chronic respiratory disease (Chap. 8).

SUMMARY

Attention has been directed to special considerations in the management of surgical diseases as they appear in the allergic infant and child. An attempt has been made to indicate when and how the surgeon may contribute.

A majority of the problems occur in asthmatics when the surgeon is mainly concerned with management of complications of the disease such as extrapulmonary air leaks, respiratory failure, intrapulmonary air loss (atelectasis, etc.), infection, bronchial obstruction and thoracic cage deformities. Indications, techniques and results of various surgical procedures have been reviewed as they would be modified when applied to problems in the allergic infant or child. These procedures include mediastinal aspiration by needle or catheter, cervical mediastinotomy, closed intercostal tube thoracostomy, tracheostomy, tracheal aspiration via direct laryngoscopy, bronchoscopy, bronchography, tracheobronchial lavage, right middle lobectomy, and open lung biopsy.

Nonallergic causes for wheezing and allergic causes for abdominal pain, intestinal malfunction and bleeding contribute further to the surgeon's involvement. The expected increase in morbidity that occurs in the pediatric surgical patient who also has obstructive pulmonary disease can be minimized by careful surgical and anesthetic management.

REFERENCES

1. Phillips, E.W., and Scott, W.J.M.: The surgical treatment of bronchial asthma. *Arch Surg, 19:*1425, 1929.
2. Rienhoff, W.F., Jr., and Gay, L.N.: Treatment of intractable bronchial asthma by bilateral resection of the posterior pulmonary plexus. *Arch Surg, 37:*456, 1938.
3. Miscall, L., and Rovenstine, E.A.: The physiologic basis for the surgical treatment of asthma. *Surgery, 13:*495, 1943.
4. Nakayama, K.: *Carotid Body.* Tokyo, Gakujutsushoin Publishers, 1948.
5. Nakayama, K.: Surgical removal of carotid body for bronchial asthma. *Dis Chest, 40:*595, 1961.
6. Overholt, R.H.: Resection of carotid body for asthma. *JAMA, 180:*809, 1962.
7. Phillips, J.R.: Removal of the carotid body for asthma and emphysema. *Southern Med J, 57:*1278, 1964.
8. Committee on Therapy, American Thoracic Society: Current status of the surgical treatment of pulmonary emphysema and asthma. *Amer Rev Resp Dis, 97:*486, 1968.
9. Curran, W.S. *et al.:* Glomectomy for severe bronchial asthma. A double-blind study. *Amer Rev Tuberculosis, 93:*84, 1966.
10. Segal, M.S., and Dulfano, M.J.: Glomectomy in the treatment of chronic bronchial asthma: A report of 15 unsuccessful cases. *New Eng J Med, 272:*57, 1965.

11. Holton, P'., and Wood, J.B.: The effects of bilateral removal of the carotid bodies and denervation of the carotid sinuses in two human subjects. *J Physiol, 181:* 365, 1965.
12. Read, C.T.: Most thoracic surgeons oppose glomectomy in management of asthma. *JAMA, 191:*7, 1965.
13. Jorgensen, J.R., Falliers, C.J., and Bukantz, S.C.: Pneumothorax and mediastinal and subcutaneous emphysema in children with bronchial asthma. *Pediatrics, 31:* 824, 1963.
14. Bierman, C.W.: Pneumomediastinum and pneumothorax complicating asthma in children. *Amer J Dis Child, 114:*142, 1967.
15. McNicholl, B.: Pneumomediastinum and subcutaneous emphysema in status asthmaticus requiring surgical decompression. *Arch Dis Child, 35:*389, 1960.
16. Pecora, D.V., Yeigian, D., and Hochwald, A.: Tracheotomy in the treatment of severe mediastinal emphysema. *JAMA, 166:*354, 1958.
17. Downes, J.J., Wood, D.W., Striker, T.W., and Pittman, J.C.: Arterial blood gas and acid-base disorders in infants and children with status asthmaticus. *Pediatrics, 42:*238, 1968.
18. Richardson, W.R.: Surgical complications of staphylococcic pneumonia in infants and children. *Amer Surg, 27:*354, 1961.
19. Richardson, W.R.: Thoracic emergencies in the newborn infant. *Amer J Surg, 105:* 524, 1963.
20. Holmes, S.L., and Mark, J.B.D.: A simple apparatus for underwater seal in pneumothorax of the newborn. *J Pediat Surg, 3:*87, 1968.
21. Winter, P.M., and Lowenstein, E.: Acute respiratory failure. *Sci Amer, 221:*23, 1969.
22. Downes, J.J., Wood, D.W., Striker, T.W., and Lecks, H.I.: Diagnosis and treatment: Advances in the management of status asthmatics in children. *Pediatrics, 38:*286, 1966.
23. Richards, W., and Siegel, S.C.: Status asthmaticus. *Pediat Clin N Amer, 16:*9, 1969.
24. Downes, J.J., Wood, D.W., Striker, T.W., and Haddad, C.: Acute respiratory failure in infants with bronchiolitis. *Anesthesiology, 29:*426, 1968.
25. Striker, T.W., Stool, S.E., and Downes, J.J.: Prolonged nasotracheal intubation in infants and children. *Arch Otolaryng, 85:*210, 1967.
26. Smith, R.M.: *Anesthesia for Infants and Children,* 3rd ed. St. Louis, C.V. Mosby, 1968.
27. Melampy, C.N.: Personal communication.
28. Stool, S.E., Campbell, J.R., and Johnson, D.G.: Tracheostomy in children: The use of plastic tubes. *J Pediat Surg, 3:*402, 1968.
29. Talbert, J.L., and Haller, J.A., Jr.: Improved Silastic tracheostomy tubes for infants and young children. *J Pediat Surg, 3:*408, 1968.
30. Strick, L.: Breathing and physical fitness exercises for asthmatic children. *Pediat Clin N Amer, 16:*31, 1969.
31. Hertzler, J.H., and Chiappe, G.: Pectus excavatum, a critical evaluation of treatment. *J Mich Med Soc, 59:*890, 1960.
32. Ravitch, M.M.: Operative treatment of congenital deformities of the chest. *Amer J Surg, 101:*588, 1961.
33. Derbes, V.J., Weaver, N.K., and Cotton, A.L.: Complications of bronchial asthma. *Amer J Med Sci, 222:*88, 1951.
34. Currarino, G., and Silverman, N.: Premature obliteration of the sternal sutures and pigeon breast deformity. *Radiology, 70:*532, 1958.

35. Frazier, C.A.: Personal communication.
36. Brown, A.L.: Pectus excavatum (funnel chest) anatomic basis; surgical treatment of incipient stage in infancy; and correction of deformity in fully developed stage. *J Thoracic Surg, 9*:164, 1949.
37. Oechsli, W.R.: "Spontaneous" rib fractures in asthmatic patients. *J Thorac Surg, 5*:530, 1936.
38. Wittig, H.J., and Chang, C.H.: Bronchiolitis or asthma. *Pediat Clin N Amer, 16*: 55, 1969.
39. Wittenborg, M.H., Gyepes, M.T., and Crocker, D.: Tracheal dynamics in infants with respiratory distress, stridor and collapsing trachea. *Radiology, 88*:653, 1967.
40. Dees, S.C., and Spock, A.: Right middle lobe syndrome in children. *JAMA, 197*:8, 1966.
41. Tarnay, T.J., Wittig, H.J., Lucas, R.V., Jr., and Warden, H.E.: Chronic and recurrent atelectasis in children. *Surgery, 62*:520, 1967.
42. Wittig, H.J., and Chang, C.H.: Right middle lobe atelectasis in childhood asthma. *J Allergy, 39*:245, 1967.
43. Kjellman, B.: Bronchial asthma and recurrent pneumonia in children: Clinical evaluation of 14 children. *Acta-paediat Scand, 56*:651, 1967.
44. Lecks, H.I., and Kravis, L.P.: The allergist and the eosinophil. *Pediat Clin N Amer, 16*:125, 1969.
45. Brock, R.C., Cann, R.J., and Dickinson, J.R.: Tuberculous mediastinal lymphadenitis in childhood: Secondary effects on the lungs. *Guy Hosp Rep, 87*:295, 1937.
46. Brock, R.C.: *The Anatomy of the Bronchial Tree with Special Reference to the Surgery of Lung Abscess.* New York, Oxford University Press, 1946.
47. Graham, E.A., Burford, T.H., and Mayer, J.H.: Middle lobe syndrome. *Postgrad Med, 4*:29, 1948.
48. Schuster, S.R., Schwachmann, H., Harris, G.B.C., and Khaw, K.T.: Pulmonary surgery for cystic fibrosis. *J Thorac Cardiovasc Surg, 48*:750, 1964.
49. Kulczycki, L.L., Mueller, H., and Schwachmann, H.: Respiratory allergy in patients with cystic fibrosis. *JAMA, 175*:358, 1961.
50. Heiner, D.C., Sears, J.W., and Kniker, W.T.: Multiple precipitins to cow's milk in chronic respiratory disease. *Amer J Dis Child, 103*:634, 1962.
51. Klassen, K.P., Anlyan, A.J., and Curtis, G.M.: Biopsy of diffuse pulmonary lesions. *Arch Surg, 59*:694, 1949.
52. Gaensler, E.A., Moister, M.V.B., and Hamm, J.: Open lung biopsy in diffuse pulmonary disease. *New Eng J Med, 270*:1319, 1964.
53. Gibbon, J.H., Jr.: *Surgery of the Chest.* Philadelphia, W.B. Saunders, 1962, p. 416.
54. Manfredi, F., and Krumholz, R.: Percutaneous needle biopsy of the lung in evaluation of pulmonary disorders. *JAMA, 198*:176, 1966.
55. Gellis, S.S. *et al.*: Use of aspiration lung puncture in the diagnosis of idiopathic pulmonary hemosiderosis. *Amer J Dis Child, 85*:303, 1953.
56. Thompson, H.T., and Pryor, W.J.: Bronchial lavage in the treatment of obstructive lung disease. *Lancet, 2*:8, 1964.
57. Ramirez, R.J.: Bronchopulmonary lavage: New techniques and observations. *Dis Chest, 50*:581, 1966.
58. Dees, S.C.: Asthma in infants and young children. *JAMA, 175*:365, 1961.
59. Jackson, C.L.: Indications and technic of bronchoscopy. *J Int Coll Surg, 6*:435, 1943.

60. Speer, F., and Todd, R.H.: *The Allergic Child.* New York, Harper & Row, 1963, p. 549.
61. Gryboski, J.D.: Gastrointestinal milk allergy in infants. *Pediatrics, 40:*354, 1967.
62. Rubin, M.I.: Allergic intestinal bleeding in the newborn: Clinical syndrome. *Amer J Med Sci, 200:*385, 1940.
63. Hunt, C.E., Papermaster, T.C., Nelson, E.N., and Krivit, W.: Eosinophilic peritonitis: Report of two cases. *Lancet, 87:*473, 1967.
64. Zuelzer, W.W., and Apt, L.: Disseminated visceral lesions associated with extreme eosinophilia. Pathological and clinical observations on a syndrome of young children. *Amer J Dis Child, 78:*153, 1949.
65. Grogan, F.T., Jr.: Food allergy in children after infancy. *Pediat Clin N Amer, 16:* 217, 1969.
66. Andreson, F.F.R.: Ulcerative colitis-allergic phenomenon. *Amer J Digest Dis, 9:* 91, 1942.

Chapter 8

Drug Reactions and the Surgical Patient

John Adriani

DURING THE PAST SEVERAL DECADES the introduction of hundreds of new therapeutic agents or modified formulations of established drugs has complicated the practice of medicine by multiplying the incidence of adverse reactions and drug-induced diseases. Physicians in all specialties have broadened their spectra of drug usage. Anesthesiologists and surgeons are not exceptions. The antibiotics, blood derivatives, plasma volume expanders, hormone and enzyme preparations, various products of biological origin, various antisera, and carcinolytic agents are among the agents that have been adopted. The days when surgeons and anesthetists rarely encountered allergic responses to drugs are over. Even so, drug allergy is not a frequent or a common cause of adverse reactions in surgical patients. Preventing such an allergic reaction may be difficult because the patient allergic to the drug is not easily identified. Sometimes patients are told that the adverse reactions they have had are due to allergy when actually they are drug induced. For example, following the use of local anesthetics, adverse reactions from overdosage, diminished tolerance, or idiosyncrasy have been incorrectly ascribed to allergy. Moreover, a patient may receive a drug one or more times previous to admission to a surgical service and become sensitized without demonstrable evidence that such sensitization has occurred.

Certain allergic patients with histories of adverse reactions to drugs or with multiple allergies to pollens, food, fibers and the like, or to combinations of these agents, are placed in a high-risk category for surgery. Anesthetists, surgeons and dentists should direct special attention to them preoperatively. The history obtained should be probing in scope, particularly for the patient with a family history characterized by seasonal or perennial rhinitis, urticaria, cutaneous eruptions and asthma. In the case of adverse drug reactions, even if a careful history is taken, a subsequent reaction is difficult to predict; as a matter of fact, a history of no difficulty with a drug on prior use may be misleading. Sometimes the fact that a drug is responsible for adverse reactions is often overlooked. Onset may be characterized by fever, chills, subclinical hepatitis, lymphadenopathy, swollen joints, purpura and suppression of bone marrow activity. Today, new agents may cause atypical, bizarre manifestations of "sensitization" in an allergic patient. A recently emphasized example is the hepatitis resulting from sensitization to halothane.

116

TYPES OF ALLERGENS AND REACTIONS

The term *hypersensitivity* has the same connotation as the term *allergy*. Both terms are used interchangeably, and both refer to a state of specific hyperactivity of an immunologic nature acquired by the animal body to some substance. In most cases the substance is exogenous and foreign to the body. An allergic response results in harm rather than protection to an individual and is not physiologic. Such a response, nonetheless, is placed in the category of an immunologic response because it is due to the interaction of an antibody with an allergenic or antigenic agent.

A substance that induces a state of allergy is called an allergen. It must be differentiated from one that is immunogenic. An immunogenic substance induces formation of antibodies that produce a protective state of immunity. An allergen likewise causes antibodies to form, but these, unlike immune bodies, are not protective. Allergens are, in most cases, large, complex molecular structures of a protein or polysaccharide nature. Their molecular weight in many cases exceeds 10,000. Surgeons and anesthetists, therefore, must bear in mind that the substances derived from biologic sources, such as glandular extracts, vaccines, sera and colloidal solutions, may contain traces of foreign proteins, cause sensitization after single or repeated injections and induce an allergic state.

However, some allergens are chemical structures of lesser complexity and of a low molecular weight. They do not by themselves induce the formation of antibodies; however, they combine with a larger carrier, a protein or polysaccharide molecule within the animal body, and then act as allergens and provoke antibody formation. These fragments, called haptenes, are capable, on subsequent exposures, of reacting with the existing carrier molecules and interacting with antibodies to cause an allergic response. The large foreign molecules are capable of inducing antibody formation alone, without combination or interaction with other chemical substances. All antigens could correctly be classified as allergens, but all allergens are not necessarily antigens because they may fail to cause antibody formation in nonsusceptible individuals.

Most drugs are capable of inducing allergy, particularly in high-risk patients, although some drugs are far more allergenic than others. For example, more individuals became sensitized to penicillin than to procaine when the combination was used.

Two categories of allergen-antibody responses are recognized—the *immediate* and the *delayed*. The *immediate* allergic responses are caused by interaction of *circulating* extracellular antibodies and an antigen. The *delayed* response results from the interaction of an allergen with intracellular antibodies. The cells involved in the reaction in most cases are believed to be lymphoid cells. Injury to tissues results in susceptible organs in either case.

Volatile Anesthetics and Allergenic Reactions

Drugs Acting as Haptenes

Almost any drug, be it organic or inorganic, may act as a haptene and combine with heavy carrier molecules to form antigens. Hypnotics, narcotics, antibiotics and other drugs used by anesthetists and surgeons are no exceptions. Even finely divided metals such as copper, gold, or silver may act as haptenes. Allergic reactions to volatile anesthetic drugs are uncommon, however, because volatile anesthetics are inert and loosely bonded to cell receptors; they are therefore unlikely to act as haptenes. Reactions have been reported only rarely, and even then there was some question as to the etiology of the reaction. The bonding in these cases most likely is the Van der Waal type (electronic attraction similar to cohesion) or hydrogen bonding, neither of which forms a strong chemical bond within the body. Such bonding is readily reversible. The hepatitis following the administration of halothane, however, is thought to be due to a "sensitization reaction." Special mention will be made of this phenomenon later in this chapter.

Halogenated Hydrocarbon Anesthetics

The halogenated hydrocarbons were popular for many years as anesthetics but the majority, due to their cardiotoxic and hepatotoxic nature, have been relegated to obsolescence. The introduction of the fluorocompounds has revived interest in halogenated inhalation anesthetics. Incorporation of the fluorine atom into chlorinated and brominated hydrocarbons and ethers enhances their stability and diminishes their cardiotoxicity and hepatotoxicity in many cases. Increased awareness of, and fear of, the fire and explosion hazards in operating rooms have compelled anesthetists to promote the widespread use of these compounds in spite of drawbacks. Their chief virtue, in many cases, is that they are either nonflammable or possess a low degree of flammability. Halothane, a fluorinated, chlorinated, brominated hydrocarbon, and methoxyflurane, a chlorinated and fluorinated ether, have supplanted ether and cyclopropane as inhalation anesthetics in certain parts of the country where static electricity is a problem and fear of explosions and possible medicolegal implications are present.

The cardiotoxic effects of halogenated volatile anesthetics are pharmacologic in nature and not linked to cytotoxicity of the myocardium or the pacemaker tissues. They occur during the administration of the drug and are reversible, therefore functional.

The hepatotoxic effects, on the other hand, are delayed, often irreversible, and are due to cytotoxicity. Two types are recognized: (a) direct injury to the cell and (b) injury by hepatitis. Some halogenated hydrocarbons, particularly single carbon saturated compounds such as chloroform and carbon tetrachloride, exert a direct cytotoxic effect on the liver cells. Swelling of the liver parenchyma, followed by necrosis, is observed. Cells in the center of the liver lobule are the most susceptible. The exact causative mechanism is not known, but the available experimental data suggest that vasospasm induced by the drug reduces hepatic blood flow. The cells in the center of the lobule are most remote from the capillary bed supplying the liver and are therefore less adequate perfused and more subject to injury from anoxia. The lesions noted in man are reproducible in animals and are dose related. Whether or not sensitization is involved is not known.

The hepatotoxicity following the administration of halothane and methoxyflurane differs in many respects from that resulting from chloroform. Neither the clinical syndrome of hepatitis nor the lesions in the liver are reproducible in animals. The etiology is unknown. Some clinicians speculate that the hepatic changes are due to "sensitization." Some proof supporting this contention is available. Anesthesiologists exposed to the drug day in and day out get hepatitis when given the drug in anesthetic concentration. Hepatitis has followed the single administration of halothane, but its incidence has been greater following repeated administration. When the first administration has been followed postoperatively by unexpected and unexplained fever, the feeling prevails that the drug should not be repeated. However, fever occurs postoperatively from many causes. Determining the etiology of the hepatitis that develops has been difficult because of the presence of other incriminating factors, in these cases such as the use of blood, blood derivatives and adjunctive drugs with a hepatotoxic potential, such as phenothiazines or steroids. It is not possible to incriminate the anesthetic agent alone in most cases studied. The issues have further been clouded by a retrospective statistical study which implicates other nonhalogenated anesthetics as well as the halogenated. Thus, the answer to the question remains unresolved at the present time, and hepatitis continues to occur sporadically following the use of halothane.

Possible Role of Products of Biodegradation

It is commonly believed that most inhalation anesthetics are totally inert in the body, but recent data reveal that this is not strictly correct. Ethers and hydrocarbons labeled with ^{14}C undergo slight but insignificant degrees of biotransformation. Halothane and trichloroethylene are the two compounds of interest in this respect. After the administration of halothane, trifluoroacetic acid appears in the urine in substantial quantities and continues to be excreted for several days. Studies in animals reveal that of the total quantity of halothane inhaled, 10 percent or more is metabolized and that traces of by-products appear in the urine for four or five days after anesthesia. Similar findings have been reported in man. Trichloroethylene also undergoes significant degrees of biotransformation. The final product is trichloroacetic acid. This, together with trichloroethanol and other intermediate products, appears in the urine of patients inhaling the agent. Sensitization from trichloroethylene has not been as serious a problem as it has been with halothane.

The possibility exists that intermediate and final by-products of the biodegradation of halogenated compounds may act as haptenes and cause sensitization or even injury to the liver cells. However, antibodies have not been demonstrated with certainty in the blood of patients with hepatic injury developed following the use of these drugs. Thus, allergy in the usual sense of the term is not a problem of major concern with volatile anesthetics. The administration of volatile anesthetics, therefore, presents few problems as far as drug-induced allergies in the true sense are concerned.

Nonvolatile Anesthetics and Hypnotics and Allergic Reactions

The development of allergy with use of nonvolatile drugs, on the other hand, is markedly different. Nonvolatile drugs frequently act as haptenes and combine with proteins or polysaccharides in the plasma and the tissues to produce allergens and antibodies. Immunologic responses follow subsequent exposure to the drug. Allergy may develop after an incubation period of ten days or more when a drug is administered to high-risk patients. The allergen interacts with the antibody at the "shock organs" on subsequent exposure and an allergic response results. Generally, allergy to the offending agent does not develop.

Another important factor in dealing with allergic patients is the possibility of histamine release caused by some drugs. Nonanesthetic adjunctive drugs, e.g. the muscle relaxants, particularly tubocurarine, cause large quantities of histamine to be released from the mast cells, as do morphine and other narcotics also. Histamine-releasing drugs may cause bronchospasm, urticaria, laryngeal edema and other manifestations of allergy in allergic patients. Cholinergic drugs, such as neostigmine and hexafluorenium bromide (Mylaxen®), although not histamine releasers, frequently are used by anesthetists and surgeons to reverse neuromuscular blockade due to tubocurarine, gallamine and other nondepolarizing muscle relaxants. They may not be desirable or are even contraindicated in patients who have bronchial asthma. In addition, neostigmine causes an increase in salivation and in secretions in the pharynx, larynx and trachea. Adequate doses of atropine or other anticholinergic drugs are recommended prior to administration of neostigmine to block the parasympathetic effects. In an allergic patient this type of drug is, of course, highly undesirable.

Cutaneous Reactions Due to Bromides and Iodides

Salts of alkaloids and alkaloid-type synthetic drugs, particularly those of certain muscle relaxants, such as gallamine, are derived from hydriodic or hydrobromic acid. The iodine and bromine are converted to inorganic iodides in these compounds and may cause cutaneous rashes postoperatively. These rashes may or may not be allergic in nature. Nonvolatile drugs containing bromide or iodine, such as tribromoethanol, chloral and its derivatives and certain barbiturates, may also cause cutaneous lesions. When these drugs are metabolized, they release the bromide ion, which is excreted by the sweat glands and causes a chemical dermatitis. This chemical dermatitis is not easily differentiated from a true allergic response at times; these substances can initiate both types of response.

Allergy to Local Anesthetics

Contrary to popular belief, allergy to local anesthetics is relatively infrequent. The majority of systemic reactions are due to high plasma levels resulting from the rapid absorption of the drug or inadvertent intravascular injection perfusing susceptible organs. Local anesthetics are un-ionized at blood pH and readily penetrate cell barriers. Minute traces circulating in the blood are highly active physiologically. In other words, most reactions are due to overdosage. Overdosage causes central excitation accompanied by convulsions as an initial manifestation. This primary excitation is followed by depression of the nervous system characterized by coma and respiratory arrest. Actually, local anesthetics are central nervous system depressants and if the quantity injected is massive, convulsions may be evanescent or absent. In addition, myocardial depression, accompanied by vasodilatation and hypotension, or a combination of all these is present. While other systems are affected, the two most vulnerable to local anesthetics are the central nervous system and the cardiovascular system.

Allergy to local anesthetics is a definite clinical entity however, but its infrequent occurrence is attested by the fact that sensitization to penicillin occurred far more frequently than sensitization to procaine, with which it was combined, when the combination was widely used. Allergic responses to local anesthetics are commonly followed by urticaria, swelling of the tissues of the throat, neck and base of the tongue and larynx in the immediate type of reaction. Conjunctivitis, rhinitis and bronchospasm likewise may occur. Intense bronchospasm, the symptom of which cannot be differentiated from bronchial asthma, occurs in highly susceptible individuals, usually as a delayed reaction. The first manifestation may be urticaria, which as a rule, is followed some hours later by subcutaneous edema in the area of injection or application. A progressive generalized systemic reaction occurs with edema of the face, neck, arms and other structures. When used topically in the pharynx, puffiness of the face, swelling of the pharynx and structures of the floor of the mouth or swelling of the tongue may occur and become progressively worse. The writer has been consulted in cases of laryngeal edema of such severity that tracheal intubation or tracheotomy was required to establish an airway. Convulsions, unconsciousness, hypotension and arrhythmias, which are characteristic of overdosage, are rarely, if ever, manifestations of allergy.

Cross-sensitization to drugs chemically or pharmacologically related may develop. In other words, if a patient becomes sensitized to one drug in a chemical family, sensitization may develop when a drug closely allied chemically is administered. For example, benzocaine, the ethyl ester of para-aminobenzoic acid, may induce an allergic state if used repeatedly on the skin or mucous membranes; then procaine, which is chemically allied to benzocaine, may cause an allergic response if administered without any previous exposure at a later date. Important examples of cross-sensitization to drugs are too numerous to list here.

Allergy to Adjunctive Drugs and Preparations

Dyes and Contrast Media

Allergic responses to various iodine-containing contrast media used to visualize the vascular system, the subarachnoid space, the hollow viscera, the tracheobronchial tree or the genitourinary tract may occur when diagnostic procedures are being performed during anesthesia. Actually, allergy is the least frequent of the causative factors which precipitate adverse reactions when contrast media are used. Bronchospasm is caused by local stimulation or irritation, particularly when the agent is introduced topically into the tracheobronchial tree. The instillation of iodized oils into the tracheobronchial tree often causes respiratory distress. Invariably this is due to the aggravation of an already existing decrease in pulmonary reserve by the oil coating the alveoli or by the bronchospasm precipitated by the agent. Respiratory distress from embolization, vasospasm, speed shock and other symptoms due to technical factors are more often responsible for adverse reactions than allergy. Syncope may result from the rapid injection of the contrast media. In preparation for visualizing the great blood vessels and the various vascular beds, particularly the coronary artery system or other chambers of the heart, the contrast media displaces the blood from the coronary vascular bed. A diseased myocardium may be unable to withstand the resulting period of anoxia, brief as it may be. However, allergic responses may occur, particularly after repeated examinations have been performed by using the same contrast media at intervals of several weeks or more.

Dyes used for studying hepatic and renal function or cardiac output, particularly those containing iodine and bromine, may cause allergic responses, particularly when they have been used on repeated occasions. Cutaneous reactions may occur, which must be differentiated from chemical dermatitis.

Substances Derived from Biologic Sources

Surgeons frequently use substances of biologic origin which cause sensitization and allergy. The dextrans, which are derived by the accretion of glucose molecules by means of a biologic process, contain antigenic foreign protein as a contaminant. Allergic responses have resulted. Various enzyme preparations used for dissolving necrotic tissues or breaking barriers of connective tissue, such as streptokinase, papase and hyaluronidase, or substances used to neutralize the antibodies in blood (Witebsky substance), have caused sensitization and allergic responses when administered at a later date. Patients whose blood is not type O but is transfused with high titre type O blood may be sensitized and manifest allergic responses. The white blood cells and platelets in blood likewise are responsible for allergic reactions. The unexplained chills and fever which often follow the administration of blood, even though matching tests reveal no incompatabilities, may be due to these substances.

GENERAL CONSIDERATIONS AND PRECAUTIONS

Both the surgeon and the anesthetist, therefore, must be aware of all past and present drug therapy. A competent, conscientious anesthetist interviews and examines the patient the day before the operation to evaluate the physical status from the standpoint of anesthesia. He may be aided by the recorded details of the history and physical examination on the patient's record. The examination at the time of the preoperative visit should be detailed. The history of drug intake is of utmost importance, particularly from the standpoint of drug interactions. Recurring rhinitis, sinusitis, frequent headaches, laryngeal edema, asthma, chronic cough, urticaria, eczema and other cutaneous rashes, and vague, nonidentifiable gastrointestinal disturbances after the ingestion of food or drugs are details worthy of note and evaluation. Joint pain, other manifestations of serum sickness after immunizations, transfusions and administration of biological products of a protein nature are important and should be noted. The history of reactions to blood products, plasma volume expanders, untoward responses after hypnotics, narcotics and other drugs is of interest since these substances may be used

during surgery. The history of jaundice following the use of anabolic steroids, phenothiazines, urocosuric agents and other drugs is important.

Although the history is of prime importance, one must remember that
patients are either misinformed or know little or nothing about drugs they
have taken. Furthermore, after an adverse reaction some patients have not
been apprized of its cause because the physician himself was uncertain of its
etiology. Also, physicians themselves may inadvertently misinform patients
because at times they are unable to distinguish an allergic reaction from intolerance, idiosyncrasy, or overdosage. Often patients have forgotten that
they experienced a reaction after taking a certain drug, and the history is
incomplete or they are not aware of the identity of the drug because labeling is omitted from the prescription. Sometimes sensitization is due not to
the active ingredient in a formulation, but to a binder, excipient, perservative,
or other added nontherapeutic substance. A contaminant may cause an adverse response, particularly when drugs of inferior quality which do not
meet United States Pharmacopoeia or National Formulary standards are
used. However, drugs prepared by reputable manufacturers may inadvertently be contaminated. For example, penicillin contamination has long vex-
ed drug manufacturers. Allergic responses have occurred, not to the drug
itself, but to contamination from penicillin. Allergy may also be due to the
breakdown products in outdated drugs instead of the pure ingredient. Thus,
the medical and drug history does not necessarily reveal all essential data.

SYSTEMIC ANAPHYLACTIC REACTIONS

An anaphylactic reaction differs from other allergic ones in a number of
ways. An anaphylactic reaction is characterized by the release of highly active
pharmacologic materials in large quantities from certain body stores, principally the mast cells or "target cells." Two compounds whose identities have
been established are histamine and serotoin. Presumably, an allergen or a
complex composed of an antigen which has united with an antibody acts on
platelets or interacts with precursors of proteolytic enzymes in the plasma.
These substances unite with the complement normally in the blood and
cause injury to the mast cells which, in turn, release histamine, serotonin, or
other offending agents. The released histamine injures secondary receptor
sites, usually the smooth muscle-containing organs and the vascular tissues or
glandular organs. Anaphylaxis is not necessarily a true immune response because it is not a protective activity. The patient undergoes a series of severe
convulsions within seconds of receiving the offending substance. The reaction is usually fatal within a few minutes. If he recovers he is anergic and
permanently desensitized and may then be given a dose many times greater
than the "reaction-producing" dose that ordinarily is lethal. Pulmonary em-

physema, an enlarged heart and other signs of circulatory collapse are common in patients dying from anaphylactic reaction.

Ample evidence exists that true anaphylaxis does occur in man. Some clinicians make a distinction between anaphylaxis as it occurs in animals and that in man. They refer to the human type of response as anaphylactoid. This term, however, should be applied only to reaction of an immunologic type which simulates a true anaphylactic reaction. Anaphylaxis can be antagonized or prevented by drugs such as the antihistamines and antiserotonins.

The human bronchiolar tissue may be responsive to bronchoconstrictor substances. Such substances have been isolated in the lungs of pollensensitive individuals. The response caused by this material, termed a *slow-reacting substance* (SRS-A), is not antagonized by antihistamines. It may be either entirely responsible for, or contribute partly to, the bronchospasm which occurs during systemic anaphylaxis in man. Cardiac arrhythmias or the release of plasma kinins which are potent vasodilators may be involved in causing the profound "vascular collapse" that characterizes some anaphylactic reactions.

Types of Drugs Causing Anaphylaxis

Agents that may induce systemic anaphylactic reactions in man are proteins, polysaccharides and haptenes of low molecular weight. Among protein substances used in surgery are horse antisera, various hormones such as insulin and ACTH, pituitary extract, various enzyme preparations such as chymotrypsin, trypsin, and penicillinase. The polysaccharides that have caused anaphylaxis are acacia and dextran. The haptenes, chiefly drugs, may combine with proteins or polyscaccharides to form allergens. Sodium dehydrochloate, thiamine, sulfanilamide, procaine and various other local anesthetics, certain salicylates, iodinated organic contrast media, numerous antibiotics and aminopyrine are among many of the substances which may act as haptenes and cause anaphylaxis.

The volatile anesthetic drugs are rarely antigenic because their mode of bonding is weak. One therefore doubts seriously that the occasional rare report in medical journals incriminating volatile anesthetics is a description of an allergic response to these agents. Sudden death during anesthesia is usually due to overdose, anoxia, or causes not related to anesthesia. Now that evidence has been brought forth that some breakdown of volatile anesthetics occurs, the argument that metabolites may act as haptenes is valid.

Systemic and Local Types of Anaphylactic Reactions

An anaphylactic reaction need not be systemic, however. Local cutaneous anaphylactic reactions may occur. Since individual cells are injured in an anaphylactic reaction rather than large groups (tissues), histologic changes at the injection site are usually absent. Transient edema and hyperemia may be present. General anesthesia does not protect against anaphylaxis or systemic allergic reactions; therefore, fatal anaphylactic reactions or allergic response may occur in anesthetized patients or develop after anesthesia.

The Clinical Picture

The clinical syndrome that is noted most frequently in a systemic anaphylactic or an anaphylactoid reaction in man is one of acute respiratory distress due to bronchospasm and angioneurotic edema involving the larynx. Both cause airway obstruction and lead to asphyxia. Diffuse erythema, urticaria, pruritus, vomiting, abdominal cramps and bloody diarrhea are additional manifestations which occur. No two pictures are identical. Severe vascular collapse characterized by syncope without preceding respiratory distress may occur in acute situations. These terminate fatally before treatment can be instituted.

Prophylaxis and treatment of anaphylaxis is discussed in Chapter 1.

TESTING FOR ALLERGY

The only method for determining with any degree of certainty whether or not a patient is allergic is to administer a test dose of the suspected drug. This procedure obviously is hazardous and places the patient in jeopardy, since the untoward response the physician is attempting to avoid may be elicited. The response may be anaphylactic or anaphplactoid in nature. Such reactions may be fatal. The dose used to perform the test could be fatal. Thus, the physician is faced with the paradoxical situation of causing the death of a patient to determine if an allergy is present.

The seemingly ideal technique for detecting drug allergy would be to employ a test for antibodies *in vitro*. Such tests could be performed in a laboratory without involving the patient. The detection of antibodies by tests of this type, even though they are complex and time-consuming, would be, relatively speaking, simpler. Unfortunately, however, none is available that is significant and of practical value. One chief objection is that correlation of the results *in vitro* with possible allergic reactions *in vivo* would be difficult. A positive response in a test tube is not necessarily assurance that a patient with suspected allergy will react adversely should the substance be administered.

When there is no choice but to use the drug that is suspected of causing an allergic response, the subject should first be desensitized. The drug or allergen is administered intracutaneously, then subcutaneously and, finally, intramuscularly, in increasing doses. The suspected agent should be administered in the distal portion of an extremity (hand or foot), so that a tourniquet may be applied and epinephrine injected immediately in the event alarming symptoms develop. The initial dose should be a diluted mixture which contains only a fraction of the contemplated therapeutic dose. When the route of administration is changed from the dermal with slow absorption to the subcutaneous with more rapid absorption, the initial dose should be less than final intracutaneous dose. To minimize the possibility of accumulation of antigen, the intervals between the injection should not be less than twenty, and close to thirty minutes. Desensitization is accomplished by the production of neutralizing or blocking antibodies by the individual.

EFFECTS OF ALLERGENS UPON THE CELLS

Many different techniques have been described for identifying drug-induced allergies that affect cells. In most cases the cellular elements involved are those of the blood, namely, the platelets and leukocytes. Allergies caused by thrombocytopenia-producing drugs such as apronalide, quinidine and quinine may be detected in this manner. Antibodies that cause platelets to disintegrate can be isolated from blood after a patient ingests apronalide. It is possible, however, that drug-induced anemias or a reduction of the platelet count may be caused by some mechanism other than allergy. Folic acid, 6-glucophosphodehydrogenates and other substances may be the causative factors rather than reactions to allergen-antibody complexes. Antibodies are absent from the blood in these cases.

CHOICE OF ANESTHESIA FOR THE ALLERGIC PATIENT

Drug Interaction and Anesthesia

The relationship of drug interactions and allergic manifestations is of vital importance when using both volatile and nonvolatile anesthetic drugs. It is common in present-day medical practice for internists, surgeons and other specialists to prescribe multiple drugs or drug combinations for both short-term and long-term therapy. Patients may still be under treatment for nonsurgical disease at the time of admission to a hospital for an operation. Cardiac glycosides, diuretics, monamine oxidase inhibitors, psychosedatives, hypoglycemic agents, antibiotics, uricosuric agents, analgesics and central-acting muscle relaxants are some of the more commonly prescribed drugs. Antibiotics such as erythromycin, anabolic steroids, phenothiazines, monamine oxidase inhibitors and uricosuric agents may cause hepatic dysfunction or have a potential for being hepatotoxic. Some consider liver or renal injury to be due to sensitization to the drug or its metabolite. Prior therapy with such potentially hepatotoxic drugs may be sufficient cause to select an anesthetic other than halothane, methoxyflurane, or other volatile anesthetics suspected of being hepatotoxic. Although no evidence exists to support the contention that subclinical hepatotoxicity from drugs used for prior therapy may be aggravated by halogenated anesthetics, nonetheless, this possibility must be entertained. The decision to use or not to use a particular anesthetic because an interaction may occur rests with the anesthetist. The history of prior or continuous therapy with all drugs is, therefore, of importance to both the anesthetist and surgeon.

Patients who have had ocular diseases and have been treated with cholinergic or anticholinergic drugs frequently become sensitized to these agents. They may cause an allergic response when they are administered for prophylaxis prior to anesthesia. Cholinergics, particularly the anticholinesterases, such as neostigmine, are used by anesthetists to reverse neuromuscular blockade caused by tubocurarine, gallamine and other nondepolarizing muscle relaxants. Anticholinesterases augment bronchospasm by their parasympathomimetic effects. Bronchospasm is further augmented in the event the tubocurarine releases histamine. They are, as a rule, contraindicated due to their bronchospasmogenic potential, particularly in patients who have recurrent bouts of bronchial asthma. Neostigmine and other cholinergic drugs also cause an increase in secretions of the salivary glands and in the oronasal pharynx unless atropine is administered to block their parasympathetic effects five to ten minutes beforehand. The use of atropine is, of course, mandatory in the allergic patient subject to bronchial asthma who receives cholinergic drugs.

Allergic Reactions During Anesthesia

It is well documented that general anesthesia does not prevent an allergic reaction when a sensitized patient is exposed to an allergen. It merely masks some of the symptoms. Symptoms such as urticaria and hypotension may appear during anesthesia while some may be suppressed but appear later when the patient regains consciousness and reflex activity.

Anesthetists and surgeons often overlook the fact that allergic reactions, particularly the cutaneous type or those involving the mucous membranes, may be caused by such auxiliary material as plastic substances, rubber masks, mask retainers, head bands, adhesive, intratracheal catheters and chemical agents and soaps used for sterilizing and cleansing these items (see Chap. 3).

Selection of Anesthesia for Allergic Patients

The selection of anesthesia for the allergic patient, as a rule, presents few or no difficulties if the patient is asymptomatic. In most cases the choice is immaterial and can be based upon the preference of the anesthetist. The two allergic conditions most commonly encountered and of the most concern to the anesthetist are bronchial asthma and allergic rhinitis (hay fever, sinusitis). The latter presents difficulties principally when secretions are present and the nasal passages are edematous and occluded, or when "post-nasal drip" is present. Premedication with an antihistamine and an anticholinergic drug is not only highly desirable but is strongly recommended to obviate the collection of secretions.

The selection of anesthesia in a patient with asthma has been a subject of considerable discourse and one which has prompted publication of innumerable papers. No rule of thumb can be formulated in the selection of the proper method since each asthmatic patient behaves differently. Evaluation must be made on an individual basis. Drugs that cause laryngospasm or bronchospasm should be avoided in these patients. Induction of anesthesia using high concentrations of volatile anesthetics which are pungent may initiate laryngeal and bronchial spasm. Some anesthetic vapors are more pungent than others and vary in spasmogenic activity more than others. The pungency of the vapors decreases in the following order: ethyl ether, vinyl ether, chloroform, trichloroethylene, halothane, methoxyflurane, cyclopropane, ethylene and nitrous oxide. Halothane and methoxyflurane, like ether, are bronchodilators in anesthetic concentrations. An "asthmatic attack" may develop, however, while any inhalation anesthetic agent is being administered, particularly during the induction period. Agents of the bronchodilator type, such as ether or halothane, break the spasm as anesthesia is deepened.

Ether in anesthetic concentrations circulating in the blood has been used in the relief of "status asthmaticus." Inhaled concentrations required for induction are laryngo-spasmogenic and bronchospasmogenic. The concentration must be increased gradually to avoid initiating an "asthmatic attack." Halothane vapor is less irritating to the upper respiratory passages in anesthetic concentrations than the other vapors. Its vapors do have some degree of pungency and may also initiate spasm in susceptible individuals, in spite of statements to the contrary. No inhalation anesthetic is immune. Cyclopropane and thiopental also are bronchoconstrictors and have spasmogenic qualities. Whether or not they may precipitate bronchospasm in an asthmatic patient is difficut to predict. Therefore, these agents are avoided whenever possible in patients with bronchial asthma. However, the contraindication is relative, and dogmatic statement that they should not be used is subject to modification when each patient is considered on individual merits. Cyclopropane is an ideal agent in asthmatic patients who are symptom-free. These agents may be used in cases of allergic rhinitis.

The combination of thiopental or other ultra short-acting barbiturate and a muscle relaxant often used by anesthetists in the so-called crash intubation technique may precipitate bronchospasm because these drugs do not dull the tracheobronchial reflexes. As a matter of fact, thiopental augments this type of activity, and there is a greater tendency toward development of bronchospasm. The addition of nitrous oxide may partly obtund the reflex activity, but it does not completely abolish it; therefore, this triple combination, which is widely used by many anesthetists, is not necessarily a good choice in patients with a history of asthma.

Evaluation of Allergic Patients for Operation

Anesthetists are particularly interested in the status of the upper and lower respiratory tract in the management of allergic patients. Pulmonary function studies are of importance and require careful study (Chap. 4). Patients who have recurrent rhinitis and sinusitis should be examined carefully because many of them have chronic or secondary infection, excess lymphoid tissue, polyps, deviations of the nasal septa, edema and thickening of the membranes in the nasal passage and other portions of the upper respiratory tract. All of these cause partial airway obstruction or predispose to trauma during nasal and oral intubation or the insertion of oropharyngeal airways. Unless secretions are controlled with atropine or scopolamine during anesthesia, they may cause obstruction if excessive, or even precipitate laryngospasm and bronchospasm. Atelectasis may develop postoperatively. The higher-than-normal incidence of postoperative respiratory complications occurs in patients who have upper respiratory tract infections or excessive secretions preoperatively. The larynx of allergic patients may be inflamed or edematous. These states may be further aggravated by intratracheal intubation. Lubricants used to facilitate insertion of airways containing local anesthetics, such as benzocaine, lidocaine, or tetracaine, may aggravate the inflammation. A patient may be allergic to these substances as well as the lubricating base, preservative, or other ingredients in the preparation, and systemic symptoms of allergy may occur. The anesthetist must know whether he is dealing with uncomplicated "pure" bronchial asthma or a mixed condition in which bronchitis or other obstructive and inflammatory disease coexists with asthma. Differentiation between "cardiac" and allergic asthma must be made. Each has an entirely different etiology and presents different symptoms and problems in anesthetic management.

Premedication of Allergic Patients

The purpose of premedication is threefold: (a) to allay apprehension; (b) to produce an additive effect with impotent agents (nitrous oxide); and (c) to provide prophylaxis. Sedation to allay apprehension is essential, particularly in asthmatic patients, since a psychogenic component of the disease plays a role in the development of an attack. As a general rule, narcotics are best omitted because they depress the respiration, release histamine, and have a spasmogenic effect on smooth muscle. Barbiturates may be used instead. An anticholinergic, atropine, or scopolamine is necessary to inhibit formation of secretions. Scopolamine is preferred because it enhances the sedative effect of narcotics and barbiturates and acts additively with the narcotic to allay anxiety as well as diminish secretions. Antihistamines are used, particularly in patients who have upper nasopharyngeal manifestations of allergy, and if they have a sedative effect also act additively with barbiturates and narcotics. They have doubtful prophylactic value in patients with asthma, and whether they are effective in overcoming the massive release of histamine in an anaphylactoid type of reaction is debatable. Results, by and large, have been disappointing. If steroids have been used for prior therapy, they should be continued in the usual dosage; an intravenous preparation should be available in the event hypotension occurs during operation.

DRUG REACTIONS OF ALLERGIC ORIGIN

Reaction to a drug can involve any one of the organ systems and are usually so classified. (a) General systemic reactions including both immediate and delayed effects are manifested by fever, swelling of the tissues and pain. Among these reactions are serum sickness, systemic lupus and periarteritis nodosa. With the exception of serum sickness, which is common and generally benign, these are extremely rare, fortunately. (b) Respiratory reactions are manifested by tracheitis, bronchitis, asthma and pulmonary infiltration with eosinophilia. These are common and troublesome. (c) Hematologic reactions are manifested by hemolytic anemia, granulocytosis, thrombocytopenia, eosinophilial lymphocytosis, leukocytosis and aplastic anemia. They are rare but serious, and in many cases fatal. (d) Dermatological responses may be morbilliform, macular, papular, vesicular, or bullous. These are common and troublesome and tend to be chronic and recurrent. (e) Neurologic reactions are characterized by paresthesias, peripheral neuritis, deafness, Parkinsonism, orthostatic hypotension and convulsions. They are not common. (b) Gastrointestinal. These are characterized by stomatitis, nausea, vomiting, abdominal pain, cramps, diarrhea, jaundice and pruritus. (g) Renal. These are characterized by albuminura, hematuria, hemoglobinuria, or acute renal failure. Fortunately, they are not common. (h) Cardiovascular. These are not common but are characterized by urticaria, angioedema, phlebitis, tachycardia, auricular fibrillation, angina, anaphylaxis and periarteritis nodosa.

Treatment of Reactions

The first and most important step in the management of a drug reaction is to stop the medication promptly. If it has been given intramuscularly, subcutaneously, or orally, the reaction may be delayed in onset but sustained. It is important to recognize the nature of the adverse reaction and determine its cause. Most often, withdrawal of the drug and symptomatic treatment is all that is necessary. Sometimes the usefulness of a particular drug for treating an allergy must be carefully balanced against the nature and severity of a possible reaction following administration. If a reaction occurs, a tourniquet should be applied proximal to the site of injection of the allergenic agent if it has been given into an extremity.

The treatment of allergic reactions is largely symptomatic. The conventional routine consists of administering epinephrine, an antihistamine and a steroid. In shock-like states when respiration is depressed, oxygen with intermittent positive pressure or assisted respiration is indicated, along with fluids and vasopressors. Syncope and cardiac arrest are symptoms of the anaphylactoid type of response. Treatment must be instituted immediately (see Chap. 3).

Serum Sickness

Serum sickness is frequently encountered in surgical patients following the use of antibiotics, the injection of various sera, antioxins, enzyme preparations, blood products and other protein-containing material. Antihistamines, especially, such as tripelennamine (Pyribenzamine), 10 to 20 mg, or diphenhydramine (Benadryl) every four hours are often effective. The latter is a much milder antihistamine. Epinephrine 1:1000 in increments of 3 cc subcutaneously may be used in the severe reaction accompanied by fever and edema. Epinephrine (1:500 in 1 cc oil) subcutaneously every eight hours may be used in protracted cases when a prolonged effect is desired. Severe reactions may require the concomitant administration of steroids. A large initial dose of hydrocortisone or prednisone (20 to 40 mg) is given, followed by 10 mg increments every four hours until relief occurs. The dose may then be reduced to 5 mg four times a day and gradually withdrawn over a period of seven or eight days. Prolonged reactions such as those that may follow penicillin administration are treated by using a suitable antihistamine. The physician has a long list of available preparations.

Asthma

Asthma may occur during induction or maintenance, at the conclusion of anesthesia, or postoperatively. Should it occur during anesthesia, an agent with bronchodilating effect, such as ether or halothane, may be used to relax the spasm. In postoperative patients epinephrine, 0.3 cc of 1:1000, may be administered as necessary. Mixtures containing ephedrine, aminophylline and barbiturate may be given orally if the patient is able to take oral medication. Certain fixed-ratio combinations containing ephedrine, aminophylline, potassium iodide and phenobarbital are used, but it is preferable to vary the amount of each ingredient to suit a patient's needs. Steroids such as prednisone, 10 mg every four hours, may be used in refractory cases. Chronically ill patients with recurrent asthma who have been receiving long-term therapy with steroids may experience hypotension when placed under the stress of surgery and anesthesia. Their steroid therapy should be continued up to the time of operation. Should hypotension develop during the operation, hydrocortisone, 100 to 200 mg IV, may be administered.

Reactions to Protein Material

Antihistamines will, as a rule, relieve the majority of allergic manifestations caused by endocrine and other products of biologic derivation. The majority of reactions are urticarial or pruritic in nature, due to the proteins in the preparations. As is the case with other agents, the physician should be familiar with several antihistamine drugs and determine by trial and error which is best. For example, tripelennamine hydrochloride, 50 mg, chlorpheniramine maleate (Chlor-Trimeton®) 5 mg or diphenhydramine hydrochloride (Benadryl), 25 to 30 mg, may be prescribed orally every four hours. Cold compresses, calamine lotion, or dilute solutions of baking soda may be used locally for symptomatic relief or at least for the placebo effect. Ephedrine sulfate, 25 mg, acts as an antidecongestant and may reduce nasal obstruction. This may be prescribed together with an antihistamine. Should a reaction characterized by bronchospasm occur during anesthesia, ephedrine, or other bronchodilating drugs such as isoproterenol may be used. In conscious patients isoproterenol may be given sublingually. In severe reactions or shock epinephrine 1:1000 may be administered subcutaneously. Vasopressors such as levarterenol and mephentermine sulfate should be used when hypotension is not overcome by epinephrine. If the patient does not respond to these agents, steroids may be used. Hydrocortisone, 100 mg intravenously, may be used in urgent situations. Usually, orally administered steroids are adequate. One may commence treatment with prednisone (40 to 80 mg) daily. The response to steroids is not immediate.

Reactions to Transfusions

Some transfusion reactions may be allergic in nature, in which case transient urticaria is the most common manifestation. It is important to remember that urticaria is a serious sign since it may herald the onset of a hemolytic reaction due to incompatibility or mismatching of blood. The absence of hemoglobin from the plasma layer of a centrifuged specimen of blood will assist in making the differential diagnosis. Simple allergic responses are relieved by epinephrine and antihistamines. Steroids are seldom required.

Certain patients develop chills, fever, and, at times, urticarial reactions which are ascribed to proteins from disintegrated leukocytes and platelets Antihistamines and sedatives and temperature-reducing measures are effective. The routine addition of drugs, particularly the antihistamines, to blood for prophylactic purposes is strongly discouraged. They may mask the early onset of symptoms of a major reaction. Furthermore, antihistamines are vasodilators that may cause hypotension, which confuses the clinical picture. The use of steroids to prevent a hemolytic reaction after incorrectly typed and matched blood is given or to prevent anaphylaxis has no clinical basis.

Bacteria may, under certain circumstances, grow in blood. When contaminated blood is inadvertently infused, severe anaphylactoid reactions invariably occur. Urticaria, chills, high fever and severe shock are the usual manifestations of such a mishap. Vasopressors to maintain the blood pressure, antihistamines and large doses of steroids such as hydrocortisone, 100 mg IV every four hours, are indicated under these circumstances. In addition, appropriate antibiotics and chemotherapeutic agents should be given to overcome septicemia.

REFERENCES

1. Adriani, J.: *Appraisal of Current Concepts in Anesthesiology.* St. Louis, Mosby, 1968, vol. 4, pp. 84-98.
2. Austin, K.F.: Systemic anaphylaxis in man. *JAMA, 192:*108, 1965.
3. Austin, K.F., and Humphrey, J.H.: *In vitro* studies on the mechanism of anaphylaxis. *Advances Immun, 3:*1, 1963.
4. Budd, M.A., Parker, C.W., and Narden, C.W.: Evaluation of intradermal skin tests in penicillin hypersensitivity. *JAMA, 190:*203, 1964.
5. Chelley, W.B.: Further experiences with indirect basophil tests. *Arch Derm, 91:* 165, 1965.
6. Crowle, J.A.: *Delayed Hypersensitivity in Health and Disease.* Springfield, Charles C Thomas, 1962.
7. Harris, M.C., and Shure, N.: Sudden death due to allergy tests. *J Allerg, 21:*208, 1950.
8. Hunt, E.L.: Death from allergic shock. *New Eng J Med, 228:*502, 1943.
9. James, L.J., and Austin, K.F.: Systemic anaphylaxis in man. *New Eng J Med, 275:* 97, 1964.
10. Smith, J.W., Johnson, J.E., and Cluff, L.E.: Studies on the epidemiology of adverse drug reactions. *New Eng J Med, 274:*998, 1966.

Chapter 9

Allergic Contact Dermatitis in Surgical Patients and Personnel

Alexander A. Fisher

BOTH THE patient undergoing surgery and surgical personnel in the course of an operative procedure are exposed to many contact allergens included in antiseptics, topical medications, local anesthetics, various adhesives and metallic and rubber commodities. A patient sensitized to such allergens may subsequently react with a contact-type dermatitis when exposed to the *systemic* administration of medications immunochemically related to these allergens.[1]

GENERAL PRINCIPLES OF MANAGEMENT

The offending contactant is identified by closely questioning the patient concerning his exposure to topical and systemic medication, before, during and following surgery. Patch tests may be necessary to help clinch the diagnosis of contact allergy but are not useful in identifying contact irritants. The patient's record listing the offending irritants and sensitizers is invaluable to the surgeon treating such a patient.

To avoid recurrences of such dermatitis, the patient should be instructed on how to obtain nonsensitizing substitutes for the offending contactant and should be warned against exposure to chemicals and medications that may "cross-react" with the specific contact allergen.

ECZEMATOUS CONTACT DERMATITIS

Topical Treatment

In the *acute stage* of eczematous contact dermatitis, characterized by redness, swelling and often blisters and weeping, cold, wet dressings of Burow's solution diluted 1:20 with ice-cold water is usually efficacious. Such dressings should not be used over too wide an area at one time. Between applications of the wet dressings, zinc oxide ointment may be applied to prevent the lesions from crusting and becoming too dry.

In the presence of secondary infection in the acute phase, a solution prepared by thoroughly dissolving a 2 grain potassium permangnate tablet in a quart of water may be used as a wet dressing. When there is persistent oozing, a 0.5% solution of silver nitrate made by adding 1 teaspoonful of 10% silver nitrate solution to a quart of water may be employed as an astringent. When the infection is widespread, confinement to bed for a few days or immobilization of a markedly edematous extremity may be indicated. As soon as vesiculation and edema begin to subside, the various corticosteroid creams and lotions may be substituted for the zinc oxide ointment.

In the *subacute* or *chronic* phase, an eruption that becomes impetiginous or otherwise infected may be treated with erythromycin ointment, an effective antibiotic preparation free of lanolin and "paraben" preservatives that may sensitize certain individuals. Intralesional injections of triamcinolone suspension (2 mg/cc) are useful therapeutic agents in persistent lichenified areas of contact dermatitis. In intertriginous areas, Castellani's carbolfuchsin paint, diluted with three parts water, may be applied, particularly when the eruption becomes chronic and is accompanied by maceration and hyperhidrosis. After the paint has dried, a corticosteroid cream may be superimposed.

Systemic Treatment: Corticosteroids

The ability of systemically administered corticosteroids to allay inflammation quickly has radically changed the outlook in severe, widespread and disabling eruptions and has greatly reduced the need for hospitalization. Provided there are no contraindications, particularly the presence of an active peptic ulcer, systemic corticosteroid therapy is indicated whenever the eruption is disseminated and disabling.

In adults, an *initial* intramuscular injection of 40 mg of triamcinolone suspension may be given. Oral therapy with corticosteroids is then begun with the administration of 60 mg of prednisone or its equivalent over a 24-hour period for two days. This dosage is gradually decreased so that by the end of the first week of therapy 30 mg of prednisone is being given daily; 20 mg is administered daily during the second week, and 10 mg daily during the third and final week of treatment. For children the doses are proportionately reduced.

Corticosteroid therapy for allergic eczematous contact dermatitis is tapered off gradually over a three-week period in order to avoid a "rebound" of the dermatitis that may occur if the therapy is stopped before the patient's own immune mechanism can take over.

Antipruritic and Sedative Therapy

In the initial phase of eczematous contact dermatitis while the corticosteroids are beginning to exert their anti-inflammatory effect or when systemic corticosteroids are not being given, sedation with phenobarbital, $\frac{1}{2}$ grain, and aspirin, 5 grains, three times daily may be helpful. For bedtime sedation, hydroxine hydrochloride (Atarax®), 25 mg, or diphenhydramine hydrochloride (Benadryl), 50 mg, may be prescribed.

CONTACT DERMATITIS FROM ANTISEPTICS

Numerous compounds commonly used as antiseptics are contraindicated for the surgical patient with a history of allergy because of their irritating or sensitizing action. The offending agents include iodine, mercury, formaldehyde and some others used in soaps and detergents.

Iodine Compounds

Iodine should not be used concurrently with ammoniated mercury since a very irritating mercury-iodide compound is formed which can cause blisters of the skin.

When tincture of iodine or even aqueous iodine solutions are used, the area treated should not be covered or bandaged unless the iodine is first removed by alcohol, since the resulting severe, irritant dermatitis may be confused with an allergic reaction or may initiate sensitization. However, if covering the area is necessary, iodine in polyvinylpyrrolidone (PVP) may be used inasmuch as iodine in PVP forms a nonirritating colloidal solution.

Iodoform (triiodomethane) and thymol iodide (dithymol diiodide) are iodine-containing compounds that are strong sensitizers, and their use should be avoided if possible. In patients already iodine-sensitive, undecoylium chloride-iodine must also be avoided.

Once sensitization has been initiated by exposure to these topically applied iodine compounds, a widespread eczematous contact-type dermatitis, often accompanied by urticaria, may follow the systemic administration of any one of numerous compounds containing iodine or its salts. Radiopaque iodized contrast media deserve special mention here since they are widely used in presurgical studies. Iodine-sensitive patients so studied suffer generalized allergic eczematous, urticarial, or pustular eruptions.[2] Surgical or laboratory personnel too who are sensitized to inorganic iodine compounds may also acquire dermatitis from simply handling or injecting these preparations.

Mercurial Antiseptics

Topical mercury compounds should not be used concurrently with inorganic iodine or sulfur preparations since strong primary irritants such as mercury-iodine or mercury-sulfide are formed and produce severe dermatitis. The organic mercurial antiseptics include merbromin (Mercurochrome®), thimerosal (Merthiolate®), nitromersol (Metaphen), and mercocresol (Mercresin®). These compounds cross-react with the inorganic mercurials that are more potent sensitizers than the organic. Allergic re-

actions to the Merthiolate (sodium ethylmercurithiosalicylate) may be due
either to the mercurial component or the thiosalicylate portion of the
compound.[3]

The inorganic germicides that may produce allergic dermatitis in
mercury-sensitive individuals are bichloride of mercury and the phenylmercuric compounds. Exposure to the phenylmercuric salts may be occult
since they are used as preservatives in many antiseptic ointments and contra-
ceptive preparations.[4]

It is noteworthy that an allergic eczematous dermatitis due to external
sensitization by a mercurial compound may undergo exacerbation or be
reproduced through the systemic administration of a diuretic containing
mercury (Mercuhydrin®). Such eruptions may become generalized and disabling and occasionally lead to exfoliative dermatitis.[5]

Finally, great care should be taken not to allow patients to come into
contact with metallic mercury from a broken thermometer; this source of
mercury sensitization has occurred several times.

Formalin

Formalin is a 37% aqueous solution of formaldehyde gas. Surgical personnel may be exposed to formaldehyde when it is used for sterilizing purposes, as a disinfectant, or as a fixing solution for tissues and for embalming
fluid. Formaldehyde is not only a powerful sensitizer but also a potent
primary irritant. A strong solution of this compound can produce necrosis
and scarring. Prolonged contact even with weak solutions may cause extreme
dryness of the skin with fissuring. The nails become discolored, soft, or
brittle, and paronychia and even suppuration of the matrix may occur.[6]

When a person is sensitized to formalin, the mere presence of formaldehyde gas in a room where a bottle of formalin has been opened, or a minute
amount of formaldehyde on thermometers, instruments, slides and biopsy
containers is sufficient to produce dermatitis. An eczematous reaction of the
hands and circum-oral area following the use of a formaldehyde-containing
toothpaste (Fig. 9-1) such as Thermodent or wearing certain wash-and-wear
and crease-proof clothing containing sufficient free formaldehyde may produce dermatitis in sensitized patients.[7]

Eczematous contact-type dermatitis medicamentosa can also result from
the ingestion of the urinary antiseptics containing methenamine which
liberate formaldehyde in an acid medium.

Antiseptics and Detergents in Soaps

Soaps containing sulfur, resorcin, mercury and benzoyl peroxide as
antiseptics may cause allergic contact dermatitis. Many popular "germici-

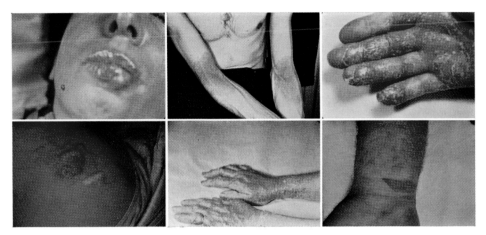

Figure 9-1. (*Top, left*) "Contact-type" dermatitis and cheilitis due to mandelamine. Allergic circumoral dermatitis due to ingestion of mandelamine in a postoperative patient who was originally sensitized to formalin in a toothpaste.

Figure 9-2. (*Top, middle*) Soap photodermatitis. Photodermatitis due to an antibacterial "scrub" soap containing tribromosalicylanilide.

Figure 9-3. (*Top, right*) Alcohol dermatitis. Rare instance of allergic dermatitis due to isopropyl alcohol with cross-reactions to other alcohols.

Figure 9-4. (*Bottom, left*) Benzocaine dermatitis. Allergic contact dermatitis due to benzocaine in a burn ointment. Patient was also allergic to procaine and paraphenylenediamine.

Figure 9-5. (*Bottom, middle*) Rubber glove dermatitis. Allergic rubber glove dermatitis in a nurse. Hypoallergenic rubber gloves are available for surgical personnel.

Figure 9-6. (*Bottom, right*) Ethylenediamine dermatitis. Patient was originally sensitized and acquired allergic dermatitis from ethylenediamine in an antifungal cream. The kodachrome shows the site of the original contact dermatitis flaring after the administration of aminophylline.

dal" soaps contain hexachlorophene (G-11). Sensitization to this compound is apparently quite rare. While hexachlorophene is not a photosensitizer, it may cross-react with certain photosensitizing chlorinated salicylanilides used as antiseptics in soap. One such photosensitizer is tribromosalicylanilide (TBS) (Fig. 9-2), a currently favorite deodorant agent in many soaps and detergents such as Lifebuoy®, Safeguard® and Praise®.[8] Foaming antiseptic detergents containing quaternary ammonium compounds are rare sensitizers.[9]

DERMATITIS FROM ALCOHOL

Alcohol not only can dry and irritate the skin but on rare occasions can be a cutaneous sensitizer. When the term *alcohol* is used in connection with its external application, the reference is usually to ethyl alcohol. When used externally, ethyl alcohol is "denatured," i.e. made unfit for drinking by the addition of certain chemicals. For industrial use, 5% methyl alcohol or acetone is often added to denature ethyl alcohol, while for ordinary or medicinal use about forty different chemicals are available for denaturing purposes. Most of these chemicals have either a bitter taste or cause emesis. Rubbing alcohol is 70% ethyl alcohol with a denaturing agent; the most popular include tartar emetic, salicylic acid, quinine sulfate, colchicum extract, brucine (an alkaloid resembling strychnine), quassin (the bitter principle of a Jamaica wood) and sucrose octa acetate (an anhydrous adhesive used in certain lacquers); they are neither potent nor frequent sensitizers. Isopropyl alcohol also, which has a slight odor resembling that of acetone, may be used as a denaturing agent. In certain hospitals, pure ethyl alcohol is used, but coloring matter such as methylene blue or amaranth pink (an azo dye which is also used to color elixir phenobarbital) is added to discourage drinking the alcohol.

Allergic contact dermatitis may, however, occasionally be caused by pure ethyl alcohol, and the allergic sensitivity usually extends to amyl, butyl, methyl and isopropyl alcohol. It may take the form of an eczematous eruption or rarely an erythematous flush or urticaria at the sites exposed (Fig. 9-3).[10] Some individuals having allergic contact sensitivity to alcohol may also react to the ingestion of alcohol with a generalized erythema.[11,12]

ALLERGIC DERMATITIS FROM LOCAL ANESTHETICS

Local anesthetics whose structure includes the para-aminobenzoic acid molecule (PABA) (Table 9-1) are common cause of allergic dermatitis. Even the slightest contact with solutions containing these agents may cause dermatitis in sensitized individuals. The marked dryness and fissuring of the dermatitis is often confined to tips of the surgeon's fingers.[13] Anesthetics

TABLE 9-1

SUBSTITUTES FOR ALLERGENIC AND TOPICAL AGENTS

Allergenic Anesthetic	*Substitute Anesthetic*
Derivatives of PABA	Not Derivatives of PABA
Butethamine	
Butacaine	Lidocaine
Procaine	Metabutoxycaine
Benzocaine	Meprylcaine
	Mepivacaine
Cross-reactors with PABA derivatives	
Derivatives of benzoic acid	
Cocaine	Lidocaine
Surfacaine®	Lidocaine
Metycaine®	Lidocaine
Topical agents chemically related to benzocaine	
Butacaine	Lidocaine
Meapaine	Lidocaine
Orthoform®	Lidocaine
Neo-Orthoform	Lidocaine

not derivatives of PABA may be used as substitutes for procaine-sensitive and benzocaine-sensitive individuals (Table 9-1). In recent years, lidocaine has received wide acceptance as a substitute for procaine. Since it is also a topical anesthetic, it may be substituted for benzocaine whenever necessary. Anesthetics derived from benzoic acid (Table 9-1) occasionally cross-react with the anesthetics based on PABA.

Benzocaine, one of the most common drugs used topically as a local anesthetic and an antipruritic agent, is a notorious sensitizer, and it and those chemically related agents (Table 9-1) should be avoided by individuals with dermatitis of the hands from any cause (Fig. 9-4). Practically all persons sensitized to benzocaine have cross-reactions with procaine. Topical lidocaine preparations may be substituted for these surface anesthetics.

Types of Dermatitis from Local Anesthetics

Surface (topical) anesthetics produce eczematous reactions while the injectable anesthetics often produce edema, erythema and urticaria in sensitized individuals. The patch test is used to prove allergic eczematous dermatitis to topical anesthetics while the intracutaneous test is used for the immediate urticarial type of reaction.[14,15]

RUBBER DERMATITIS

Surgical personnel sensitized to certain antioxidants and accelerators in rubber materials have acquired allergic dermatitis from rubber gloves, stethoscopes, rubber finger-guards, rubber bands, rubber aprons and rubber goggles. Rubber adhesive, shoes and other footwear, and rubber mammary prostheses also may produce dermatitis.[16]

Rubber Glove Dermatitis

Such dermatitis should be suspected when a dermatitis on the hand stops abruptly above the wrist. The eruption may mimic a photodermatitis. Even when the patient is not allergic to rubber, rubber gloves may make an existing hand dermatitis worse owing to an occlusive, macerating effect (Fig. 9-5).[17] For such patients and for surgeons, a neoprene* surgical glove, which, as a rule, requires less accelerators and antioxidants than do gloves made with a mixture of natural rubber, and a eudermic† surgical glove can be tolerated by many sensitized persons.

Rubber Pouches in Ileostomy Patients

Rubber pouch dermatitis should be suspected when an eruption appears under the entire disk area and on the abdomen or thigh where the rubber pouch rests. Such pouches should be replaced by all-plastic appliances.[18]

Rubber Cement Dermatitis

Certain cements employed to make an ileostomy appliance adhere to the skin may produce allergic contact dermatitis. Duo-Adhesive* may be tried as a substitute or karaya gum may be used either as a substitute for the cement, or as a protective layer to prevent skin contact with the cement. Neo-karaya, a combination of karaya with aluminum hydroxide gel, is some-what more efficient than is the combination of karaya with water in allaying inflammation and in forming a protective coating to help "insulate" skin which has become allergic to cement. For those allergic to karaya, Orabase® (Squibb), which contains pectin, may be used.

The treatment of acute dermatitis about an ileostomy stoma is much the same as that employed for acute dermatitis in other areas. Some patients obtain quick relief from the application of aluminum hydroxide from which the supernatant liquid has been removed. Hydrocortisone, 1% in plain Lassar's paste, is usually healing. This paste may be covered with Saran Wrap® and the appliance re-applied.

Rubber Dental Sheeting

In certain oral surgical procedures, a fine rubber sheeting to isolate the work area is stretched across the face and held in place by a mechanical de-

---

*Pioneer Rubber Company, Willard, Ohio.

†B. F. Goodrich Industrial Products Co., Sundries Sales Dept., Akron, Ohio.

Johnson & Johnson.

vice. In one patient facial dermatitis had developed on two occasions from this rubber dental sheeting; spraying the cheeks with Decadron® aerosol before application of the sheeting prevented the dermatitis from occurring.

Rubber Bandages

In rubber-sensitive individuals these bandages should be replaced by those made of nonrubber stretchable material such as Spandex.

ADHESIVE-TAPE DERMATITIS

Dermatitis from adhesive tape may be due to trauma, occlusion, or an allergic reaction.[19,20] Most of these skin reactions are of a mechanical nature. One type results from the shearing stresses at the tape-skin interface when tapes are applied with a degree of tension beyond the limits of physiological skin tolerance. Redness, edema and denudation may occur. A follicular, prickly-heat–like eruption follows the mechanical plugging of the follicular and sweat gland ostia by adhesive tape. It is best treated with a simple drying lotion (calamine lotion) rather than with creams or ointments. True allergic reactions may be due to rubber compounds incorporated with the adhesive mass of some tapes; usually rubber accelerators or antioxidants are the actual culprits. However, occasionally, resins and turpentine that have been added to certain rubber masses may produce allergic dermatitis. It should be noted that patients who are allergic to rubber-mass adhesive tape may also react to Scotch Tape®.

Acrylate-mass adhesive tapes,* which employ a synthetic acrylate polymer as the adherent, are desirable substitutes for the rubber-mass tapes. They may be safely used by those who are allergic to, or irritated by, the older tapes. Thus far, I have encountered no instances of allergic dermatitis due to acrylate adhesive mass itself. In addition, nonspecific, mechanical and traumatic adhesive-tape dermatitis is less likely to occur.

DERMATITIS DUE TO METALLIC INSTRUMENTS AND FOREIGN BODIES

Nickel-Plated Instruments

Sweat can readily leach out sufficient nickel from nickel-plated instruments and other objects used in surgery to produce allergic dermatitis in nickel-sensitive individuals. The same phenomenon can occur from chrome-plated intsruments, most of which contain sufficient nickel to produce dermatitis. Also, many of the needles and cannulae used for routine intravenous infusions are either nickel-plated or may contain a nickel-containing

---

*Dermacel (Johnson & Johnson) ; Blenderm and Microspore Surgical Tape (3M) .

alloy. Syncope and dermatitis in nickel-sensitive individuals following infusions from such cannulae have been reported.[21] It should be emphasized that while nickel-sensitive individuals can acquire dermatitis from both nickel-plated and chrome-plated objects, chromate-sensitive individuals can handle chrome-plated objects without difficulty provided that they are not nickel-sensitive.[22]

Foreign Bodies

The steel nails, splints and screws used for repairing fractures usually contain nickel, chrome, molybdenum and sometimes cobalt. Allergic dermatitis from such metallic foreign bodies has been reported, principally in nickel-sensitive individuals. Such dermatitis may persist until the metallic foreign body is removed.[23] Bullets and shrapnel too may contain nickel or chromium. Such missiles, until surgically removed, may produce persistent allergic dermatitis in sensitized individuals.

REFERENCES

1. Fisher, A.A.: *Contact Dermatitis.* Philadelphia, Lea and Febiger, 1967, p. 231.
2. Shelley, W.B.: Generalized pustular psoriasis induced by potassium iodide. *JAMA, 201:*13, 1967.
3. Gaul, L.E.: Sensitizing component in thiosalicylic acid. *J Invest Derm, 31:*91, 1958.
4. Morris, G.E.: Dermatoses from phenylmercuric salts. *Arch Environ Health, 1:*53, 1960.
5. Fisher, A.A.: Recent developments in the diagnosis and management of drug eruptions. *Med Clin N Amer, 3:*787, 1959.
6. Rostenberg, A., Jr., Bairstrow, B., and Luther, T.W.: A study of eczematous sensitivity to formaldehyde. *J Invest Derm, 19:*459, 1952.
7. Hovding, G.: Free formaldehyde in textiles: Cause of contact eczema. *Acta Dermatovener, 39:*357, 1959.
8. Harber, L.C., Harris, H., and Baer, R.L.: Photoallergic contact dermatitis due to halogenated salicylanilides and related compounds. *Arch Derm, 94:*255, 1966.
9. Padnos, E., Horwitz, I.D., and Wunder, G.: Contact dermatitis complicating tracheotomy: Causative role of aqueous solution of benzalkonium chloride. *Amer J Dis Child, 109:*90, 1965.
10. Fisher, A.A.: Contact dermatitis: The noneczematous variety. *Cutis, 4:*567, 1968.
11. Wasilewski, C., Allergic contact dermatitis from isopropyl alcohol. *Arch Derm, 98:*502, 1968.
12. Hicks, R.: Ethanol, a possible allergen. *Ann Allerg, 26:*641, 1968.
13. Lane, C.G., and Luikart, R., II: Dermatitis from local anesthetics with a review of 107 cases from the literature. *JAMA, 146:*717, 1951.
14. Fisher, A.A., and Sturm, H.M.: Procaine sensitivity: The relationship of the allergic eczematous contact-type to the urticarial, anaphylactoid variety: The use of xylocaine in procaine-sensitive individuals. *Ann Allerg, 16:*593, 1958.
15. Aldrete, J.A., and Johnson, D.A.: Allergy to local anesthetics. *JAMA, 207:*356, 1969.
16. Sidi, E., and Hincky, M.: Rubber glove dermatitis. *Presse Med, 62:*1305, 1954.
17. Wilson, H.T.H.: Rubber glove dermatitis. *Brit J Med, 2:*21, 1960.

18. McNamara, R.J., and Farber, E.M.: Circumileostomy skin difficulties: A study in Great Britain and the United States. *Arch Derm, 89:*675, 1964.
19. Sidi, E., and Hincky, M.: Allergic sensitization to adhesive tape: Experimental study with a hypoallergenic adhesive tape. *J Invest Derm, 29:*81, 1957.
20. Murphy, J.C., Reif, A.E., and January, H.L.: Cutaneous hypersensitivity to adhesive and Scotch tapes. *J Invest Derm, 31:*45, 1958.
21. Stoddart, J.C.: Nickel sensitivity as a cause of infusion reactions. *Lancet, 2:*741, 1960.
22. Fisher, A.A., and Shapiro, A.: Allergic eczematous contact dermatitis due to metallic nickel. *JAMA, 161:*717, 1956.
23. Foussereau, J., and Laugier, P.: Allergic eczemas from metallic foreign bodies. *Clin Dermatologica, 52:*220, 1966.

Chapter 10

Ear, Nose, and Throat Surgery and Allergy

Robert L. Goodale

CLOSE TEAMWORK of the allergist and ear, nose, and throat surgeon is essential to the successful treatment of the allergic patient who also suffers from abnormalities of the nose, sinuses, pharynx, or nasopharynx.

If the allergist is equipped to perform detailed examination of the nose, ears, and throat, then much time will be gained as he can then screen the patient and, if indicated, refer him at once to the otolaryngologist for a complete ear, nose and throat (ENT) evaluation. Often the surgeon sees the patient first because of a nose and throat complaint. Here the surgeon also should be on the alert to the possibility of an allergic factor.

The otolaryngologist must keep in mind that the allergic patient differs from the nonallergic patient suffering from an ear, nose and throat disease in that the status asthmaticus or allergic diathesis, call it what you will, is a basic defect in the patient's structure. It will continue to be present even after surgical procedures which may be clearly indicated and properly executed. In other words, surgery is primarily for the relief of ENT disease for its own sake and secondarily to alleviate the patient's allergic disability. It is not a substitute for continuing medical treatment of the allergy.

The allergic patient by inheritance or acquisition reacts to a specific factor which he inhales or ingests or with which he comes in contact. He may be affected by one or more of these agents and to varying degrees. Also, emotional stress or strain and social and psychological maladjustments can initiate or aggravate an attack of allergy. Add to this a structural deformity of the nasal passage, a chronic hyperplastic rhinitis, a sinus that fails to drain and eventually becomes chronically infected and one has a secondary factor that increases the total disability.

In discussion the ENT surgeon's position, first of all let us say that the nose, sinuses and throat are common ground for allergy and infection. An acute attack of allergy from exposure to a specific allergen will cause a violent vasomotor reaction even in an otherwise normal upper respiratory tract. Obviously the surgeon should refrain from any operation even of a minor nature under these circumstances. However, he must examine the patient to make certain there is no abnormality such as a deviated septum, hypertrophic turbinate, nasal hyperplasia or polyp, or an obstructing or chronically infected tonsil or adenoid. If some such condition is present, he can report to the allergist that, in addition to the acute episode, there are factors which might well be considered sufficiently important to warrant correction.

The selection of the proper time for surgical interference is the joint responsibility of the allergist and the surgeon. It is certainly advisable to postpone surgery until the allergic factors have been brought under control or at least until the patient has been given the benefit of whatever therapy for the allergy is indicated. At times, complete control is not obtained, and in such cases surgery may have to be undertaken if the ENT disease is severely jeopardizing the recovery.

The association of allergy with upper respiratory infection has led to much difference of opinion as to their relative etiologic roles. Does the repetition of allergy attacks render the nose and sinuses more prone to infection, and vice versa, does chronic infection of the nose and sinuses cause allergy? From a practical point of view, I prefer to regard this complex as "allergy with superimposed infection."

When a case is referred for an ENT examination, it is helpful to the surgeon to have a transcript of the allergist's notes in order to direct the evaluation into the proper channels. It also obviates repetition of allergic procedures. I have found that a checklist helps greatly in covering the history and local physical examination.

The chief complaint should be noted in the patient's own words, and he should be encouraged to tell his whole story before being asked specific questions:

A. Obstruction to breathing
   1. When first noted
   2. Unilateral, bilateral, intermittent, or continual
   3. Acquired in coruse of nasal allergy or upper respiratory infection
   4. Traumatic

B. Nasal discharge
   1. When first noted
   2. Acquired in course of infection, allergy, or nasal trauma
   3. Unilateral, intermittent, or continual
   4. Character of discharge, anterior or postnasal, serous, mucoid, purulent, crusty, sanguinous

C. Sneezing and wheezing
   2. Aggravated by
      a. Medication (drug allergy)
      b. Infection

D. Known history of sinusitis
   Patient may be confused on this point and claims he has had sinusitis solely because he has had pain or headache. However, both these symp-

toms are of prime importance in acute or recurrent acute sinus infection. They may be absent in chronic infection.

E. Is patient infection-prone
    1. Common colds, flu
    2. Sore throats, strep throats
    3. Laryngitis

F. Pulmonary history
    1. Bronchitis
    2. Pneumonia
    3. Asthma
    4. Chronic nontuberculous pulmonary disease

G. Otitic history
    1. Pain
    2. Headache
    3. Deafness
    4. Tinnitus
    5. Vertigo
    6. Discharge
    7. Extension to mastoid
    8. Association with allergic attacks

LOCAL PHYSICIAN EXAMINATION

This is a checklist to be used in all cases—allergic or nonallergic.

A. External appearance
    1. Asymmetry of head (traumatic or developmental)
    2. Deformity of nose
    3. Unusual swelling *or*
    4. External inflammatory edema or evidence of abscess in relation to the sinuses
    5. Disturbance of orbital content
    6. Abnormally small nares
    7. Malfunction of the nasal alae
    8. Disturbances of sensation or special sense

B. Intranasal examination
    1. Nasal vestibule
        a. Obstruction due to anterior dislocation of septal cartilage or local tumors
        b. Respiratory insufficiency from poorly functioning alar cartilages
    2. Nasal mucosa
        a. Color: normally pink or unusually red from acute inflammation or pale as in allergy, dull red as in chronic rhinitis

  b. Swelling
   (1) Acute edema
   (2) Chronic hyperplasia
   (3) Polyposis
  c. Secretion: watery, mucoid, purulent, dry, or crusty
  d. Tendency to hemorrhage
  e. Unusual response to stimulation such as paroxysmal sneezing
 3. Septum
  a. Deviation
   (1) Slight irregularity without obstruction
   (2) Dislocation of septal cartilage *and/or*
   (3) Deflection of vomer and perpendicular plate of ethmoid with obstruction (often this is traumatic)
  b. Hyperplasia of septal tubercle
  c. Unusual vascularity of Kiesselbach's area with tendency to epistaxis
 4. Turbinates
  a. Lower turbinate: enlarged, atrophic
  b. Middle turbinate
   (1) Enlarged by a cystic hypertrophy, polysis, hyperplasia, malposition interfering with sinus drainage
 5. Middle meatus
  a. Obstruction from
   (1) Lateral displacement of middle turbinate
   (2) Nasal polyps or neoplasm
   (3) Mucocele of ethmoid, enlarged bullar cell
  b. Discharge
 6. Nasopharynx
  a. Obstruction
   (1) Hypertrophic posterior tips of lower and/or middle turbinates
   (2) Hypertrophic adenoids
   (3) Neoplasms, polyp, fibroma, meningocele
   (4) Atresia
  b. Discharge from paranasal sinuses
  c. Eustachian tubes (secondarily blocked and/or infected from associated inflammatory disease or by 1, 2, or 3 above)
 7. Mouth and pharynx
  a. Tongue
  b. Buccal mucosa, palate, alveolar mucosa and teeth
  c. Tonsils, lateral bands and posterior wall

8. Larynx, hypopharynx and base of tongue
    a. Paralysis of vocal cord (unilateral, bilateral; adductor or abductor paralysis)
    b. Paralysis of muscles of deglutition
    c. Benign neoplasm (poly, fibroma, epithlioma, papilloma, hemangioma)
    d. Premalignant neoplasms, leukoplakia, or hyperkeratosis
    e. Malignant neoplasms
    f. Aberrant lymphoid tissue
    g. Laryngocele
    h. Hypertropic lingual tonsil
    i. Lingual thyroid gland
    j. Inflammatory edema of larynx with or without exudate
    k. Angioneurotic edema
9. Sinuses
    a. Discharge
        (1) Unilateral or bilateral
        (2) Intermittent or constant
        (3) Serous, mucoid, purulent, or sanguinous
        (4) Color of discharge: clear, white, or yellow
        (5) Odor: none or fetid
        (6) Origin: frontal sinus, antrum, anterior ethmoid cells via middle meatus, sphenoid sinus, posterior ethmoid cells via postnasal route
    b. Associated intranasal obstruction
    c. External swelling
        (1) Local swelling over sinus
        (2) Swelling affecting adjacent structures (eyes, teeth, or facial bones)
        (3) Associated inflammation
    d. Tenderness on pressure
    e. Cranial extension of disease
        (1) Osteitis
        (2) Osteomyelitis
        (3) Intracranial extension
    f. Cysts
        (1) Serous
        (2) Mucoceles
        (3) Pyoceles
        (4) Dentigerous
        (5) Dentigenous

g. Benign neoplasms
  (1) Osteoma
  (2) Fibroma
  (3) Polyp
  (4) Fibrous dysplasia
  (5) Ossifying fibroma
  (6) Wegener's granulomatosis
h. Malignant neoplasms
i. Fistula

10. Ears
  a. External ear
    (1) Congenital abnormality
    (2) Inflammation of auricle or canal
    (3) Swelling of auricle or canal
    (4) Discharge: serous, sanguinous, purulent, caseous, or fetid
    (5) Tenderness
    (6) Postaural or preauricular tenderness, and/or swelling
    (7) Facial paralysis
    (8) Exostosis of canal
    (9) Foreign body in canal
    (10) Cerumen
  b. Middle ear
    (1) Typanic membrane: color, position, intact, or perforated, mobile or fixed.
    (2) Discharge: serous, mucoid, purulent, caseous, or sanguinous
    (3) Granulations, polyps, granuloma, hemangioma, malignancy
    (4) Necrosis of ossicles
    (5) Cholesteatoma. (Possibility of mastoiditis, acute or chronic must be considered in all inflammatory conditions of the middle ear.)
    (6) Conductive deafness
  c. Inner ear
    (1) Loss of hearing (perceptive deafness)
    (2) Vertigo (labyrinthitis)
  d. Apicitis
  e. Polyneuritis (Ramsey-Hunt syndrome)
  f. Intracranial disease: acoustic neuroma, primary cholesteatoma cerebellopontine angle tumors, extension of infection of otitis origin, meningitis, epidural, subdural and brain abscess
  g. Extension of infection to lateral sinus
  h. Audiometric testing

(1) Air conduction
(2) Bone conduction with
    (a) Masking *and/or*
    (b) Discrimination testing

X-RAY EXAMINATION

Precise information is necessary in evaluation of sinus and ear disease. It is therefore essential that the following positions be taken: lateral, Water's, Caldwell "upright fluid level," and basal for sinuses; and the Townes, Attic, Law's, Schuller's, and basal for the temporal bones.

Bacteriologic evaluation is important in all cases with an associated inflammatory process. Cultures should be planted while still wet in a meat tube and also streaked on a blood agar plate. After twenty-four hours of incubation the swab from the meat tube is again streaked on a second blood agar plate and all three cultures are incubated. At the Massachusetts Eye and Ear Infirmary this method has given 80 percent more accurate identification of pathogenic organisms than can be obtained by other techniques. If antibiotic drugs are to be given, a test for sensitivity to drugs is essential. It is now evident that many strains of coagulase positive hemolytic *Staph aureus* are resistant to certain antibiotics, and precise information on the point is helpful.

ENT DISEASE COMMONLY ASSOCIATED WITH ALLERGY

Acute Allergic Rhinitis Without Infection or Structural Abnormality

The history of repeated acute attacks of sneezing, bilateral nasal congestion, watery or mucoid nasal discharge, lachrymation, or inflamed and itching eyes when the patient is exposed to the allergen to which he is sensitized is typical of acute allergic rhinitis. Because of the nasal congestion it is possible to have acute obstruction of the paranasal sinuses causing intense headache. This is typically located in the region of the sinus itself or may radiate to the parietal and occipital areas if the ethmoid or sphenoid sinuses are involved.

The examination of the nose will show an acutely edematous, pale swelling of the mucosa of the lower and middle turbinates bilaterally. There is a profuse watery or mucoid discharge. A stained smear of the secretions, as in allergy generally, will show an increase in eosinophilic leucocytes. The sinuses by transillumination and x-ray film may be normal or at most show some retained secretion and swelling of the sinus mucosa. The ethmoid bone may show some decalcification of the cell partitions if the attack is prolonged. Usually the pharynx is not remarkable. Tonsils and adenoids are not affected by the allergy. However, if this is a case of drug allergy, angioneurotic edema of the lips, fauces and larynx may be severe and in the case of the larynx lead to dangerous respiratory obstruction necessitating tracheotomy.

Acute Allergic Rhinitis with Structural Abnormality of the Nasal Passages

The history varies from the foregoing cases in that there is also some persistent obstruction to breathing due to deviation of the septum, a hypertrophic middle or lower turbinate, a cystic enlargement of an ethmoid cell, or nasal polyps, and polyposis of the mucosa.

Any surgical intervention must wait for subsidence of the acute phase. Whether to operate at all when a septal obstruction is present depends on the degree of the deflection. The septum should be resected only when the airway is significantly reduced after shrinking the mucosa. The slight irregularities that do not alter the airflow are better left alone.

The same conservatism should apply to the hypertrophic and cystic turbinate. Because of the physiologic air-conditioning role of the lower turbinate, only in extreme cases should the surgeon attempt to trim or cauterize this structure. Nasal polypi when they are pedunculated should be removed by the nasal snare. They will not respond to medical treatment. Nasal polyposis is less amenable to surgery as the polyposis usually has a broad base of attachment to the middle turbinate; therefore merely excising the redundant tissue leaves a raw surface which will regenerate a new polypoid membrane. Patients with polyposis and cystic ethmoiditis should be operated upon only in the case of advanced ethmoid disease.

Nasal Allergy and Cysts of the Sinuses

Serous cysts of the maxillary sinus are usually "silent," but if they increase in size, cause pressure on the sinus wall or predispose to infection, they should be removed.

Mucous cysts, pyoceles, and dentigenous cysts of the maxillary sinus should be removed. The presence of cysts may cause external swelling of the face where the bony walls of the sinus involved become thin from internal pressure. As the intranasal examination may be quite normal, often the existence of a cyst can be determined only by x-ray films.

Nasal Allergy and Acute Sinusitis

This should be treated conservatively by appropriate antibiotics, antiallergy therapy, and supportive medical regimen (bed rest and fluids).

Residual undrained pus may be the result of structural obstruction at outlet of the sinus or of intrasinusal edema. Conservatism is essential. Surgical interference should be undertaken only for drainage. In the case of the frontal sinus, early external trephine is indicated when all medication fails to obtain discharge via the nasofrontal duct. In the case of maxillary sinus empyema, puncture of the intranasal antrum or irrigation through the antral ostium should be done when local medication to the ostium fails to bring relief. If the suppuration persists after puncture or irrigation, an intranasal antral window should be done to provide permanent drainage. Since this is presumably a case of obstruction alone, the sinus mucosa should return to its normal state. In the case of ethmoiditis which fails to clear up after medical treatment, x-ray films may reveal the precise area which is not drained. It has been recommended that the cell in question be uncapped without the whole labyrinth being disturbed. Generally, however, the whole labyrinth is involved to such an extent that this precision is not possible. Medical treatment, antihistamines, antibiotics, steroids, and shrinking the nasal mucosa by vasoconstrictors is the best procedure. Surgery is to be done only when the disease threatens to involve the surrounding areas.

Allergy and Chronic Sinusitis

The allergist or surgeon often finds it difficult to obtain a clear story of the onset of this condition. Allergy often coincides with the original nasal infection. The two may continue ever after as a disease complex, and the history will show that although there is definite allergy, the nose and sinuses have often or continually been the seat of a bacterial infection. The discharge is usually purulent or mucopurulent. The obstruction depends on the location of the sinusitis. These cases are not infrequently complicated by pulmonary disease such as acute or chronic, asthma, chronic bronchitis, emphysema, or bronchiectasis. Examination of the nose usually shows a mucopurulent exudate covering the lower turbinates. If there is a draining antrum, frontal sinus, or anterior ethmoid, pus will be seen under the middle turbinate. Postnasal discharge may come from the sphenoid, the posterior ethmoid, or from the other sinuses if the airflow anteriorly is blocked. Tenderness over the sinuses may be absent, but usually in the case of an acute exacerbation it would be present. Extension of sinus suppuration could involve the orbit or the anterior cranial fossa if the bony wall has been so weakened by disease that it breaks through.

The mucosa of the lower turbinate is swollen, inflamed, and thickened and does not readily respond to vasoconstrictors. It may even show some epithelial hyperplasia. In the region of the ethmoid bone the mucosa is usually polypoid, but here too there is often some hyperplastic thickening especially at the anterior tip of the middle turbinate. These polyps can be dull red and thick and may contain abscessed cavities. The septal tubercles are often hyperplastic and do not respond to vasoconstricting agents.

SURGICAL OBJECTIVES

The objective of surgery is to reestablish normal physiology of the eye, ears, nose and throat by correcting a structural defect or to alleviate or eliminate an infection. In so far as possible, the surgeon should maintain a conservative attitude, preserving those structures capable of returning to normal and excising, extirpating, or obliterating, as the case may be, only when the disease is so chronic and resistant to conservative measures that it endangers the patient's well-being. It is important to individualize the problem presented by each case. There is no computer that can do this. The relative importance of the allergic disease and the ENT process must be weighed in the light of the clinical picture and experience of the allergist and surgeon. *Primum non nocere.*

SURGICAL PROCEDURES
Correction of Nasal Obstruction
Deviated Septum

The submucous resection of the septal cartilage, vomer and perpendicular plate of the ethmoid is so well described in the ENT textbooks and is so firmly established as a standard technique that little need be stated here except that this operation may be supplemented by a nasal plastic reconstruction of the nasal vestibule when there is abnormal development of the alar cartilages.

Nasal Polyp

Removal by snare is often an office procedure. However, when the polyp formation is extensive, it is better to operate in the hospital as there may be much bleeding that requires packing, and the stress and strain on the patient is too great for ambulatory care.

Hypertrophic Turbinate

Only when this causes marked obstruction is amputation of the anterior tip of the middle turbinate or trimming of the lower margin of the lower turbinate warranted. Moderate hypertrophy is better not operated upon.

Hyperplastic Mucosa

Occasionally the mucosa of the lower turbinate is hyperplastic to the extent that it will not shrink by medication. In this case, diathermic cautery can be done as an office procedure. The needle is placed either on the surface or submucosally along the turbinate at its maximal convexity, i.e. where it approaches closest to the septum. The cauterization will destroy the mucosa locally, and the ensuing fibrosis will contract the convexity and draw the turbinal mucosa away from the septum. Similar cautery of the septal tubercle is indicated when this organ is hyperplastic.

Drainage of Retained Secretion in Sinuses

Infraction of the Anterior Tip of the Middle Turbinate

If this procedure is occluding the infundibulum, can be done easily in the office and under topical anesthesia. An elevator, such as is used in submucous resection, is slipped under the tip of the turbinate and then pressed medially until the turbinate is fractured. This should allow the retained secretion to escape if its duct or ostium is otherwise patent.

Lavage of Maxillary Sinus via the Normal Ostium

This is also an office procedure and is best done under topical anesthesia. A Van Alyea cannula is introduced through the ostium, and sterile warm saline solution is gently injected into the sinus, thereby displacing the retained secretion.

Removal of Secretion by the Proetz Displacement Method

This method is especially useful in evacuating secretion from the ethmoid sinuses.

Antrum Puncture

A sharp cannula is thrust into the antrum via the lower meatus, and the irrigating solution is forced gently into the sinus, care being taken to unblock the ostium by vasoconstrictors and also to make sure that the cannula does not fit tightly. It is wise to make several punctures or one large puncture. Suddenly increasing the intrasinal pressure can lead to shock and/or collapse.

Antrum Fenestration

The intranasal antrum window is an extension of the antrum puncture technique and entails the resection of the nasoantral wall through the lower meatus. The bone and its covering mucosa is removed from the attachment of the turbinate down to the floor of the nose and anterior-posteriorly from the ascending process of the superior maxilla to the ascending process of the palate bone. Care was exercised to avoid injuring the under surface of the turbinate itself.

Frontal Trephine

Retention of purulent secretion in the frontal sinus can lead to permanent damage to the sinus mucosa and should not be allowed to persist beyond the acute phase of the infection. In fact, there may be occasions when early trephination by burr hole is indicated, especially when there is threatened extension of infection to the surrounding structures. Trephination should avoid interference with the trochlea as persistent dyplopia could result from its dislocation. For this reason the author favors trephination through the anterior sinus wall at a point just above the brow ridge and near the midline. A temporary drainage tube should be placed in the burr hole. By removing the secretion and relieving the intrasinal pressure, the swollen mucosa which has been pushed down into the nasofrontal duct should subside and allow the duct to re-open and drain. When this has occurred, the trephine opening can be allowed to close.

Radical Surgery for Severe Chronic Disease

Extirpation of Diseased Mucosa of Antrum and Formation of Permanent Intranasal Fenestration for Drainage

This is the Caldwell-Luc operation. It may be combined with ethmoid exenteration by transantral ethmoidectomy or by intranasal ethmoidectomy (the latter can be an independent procedure) .

External Ethmoidectomy via Orbital Approach

This is an independent procedure, but if there is sphenoid sinus involvement also, it is a most satisfactory method of operating on both of these sinuses.

Frontal Sinusotomy

Many techniques are described in the textbooks. The orbital approach employs the Jansen-Lynch-Smith techinique combined with ethmosphenoidectomy or its modifications. The frontal approach is an osteoplastic technique for conservative removal of cysts, ethmoid extension into the frontal area, or osteoma.

Frontal Sinusectomy

This is for severe infected and is intended to ablate the sinus entirely or to obliterate it. By the Riedel technique, the anterior and inferior wall is removed and the mucosa extirpated. This followed by tissue-implant allowing the cavity by sclerosis or osteogenesis to obliterate itself. By the osteoplastic technique, preservation of bony structures but extirpation of mucosa is followed by tissue implant or allowing the cavity to obliterate itself by sclerosis and osteogenesis.

Nasopharyngeal Obstruction

Choanal polyp is best removed by a tonsil snare introduced through the nose and adjusted by per oral finger manipulation. It is then grasped by a tonsil tenaculum and removed through the mouth after amputation. Hypertrophic adenoid and lymphoid tissue is removed by adenotome and ring punch excision. It is best done if palate is elevated and retracted so as to give a direct view of fossa of Rosenmüller. It is important to remove accessory lymphoid tissue from the eustachian tube also. Hypertrophic posterior tips of turbinates are removed by introducing a tonsil snare through the nose and adjusting it around the tip by finger in the nasopharynx.

Chronic Tonsillitis and Hypertrophy

The total extirpation of the tonsil is best done by extracapsular dissection and snare amputation.

POSTOPERATIVE CARE

It is important to follow the surgical procedure by treating the allergic factors and controlling any bacterial complication. This will require a long-range program of checkups by both the allergist and the surgeon. Much can be gained by explaining this to the patient before surgery. He is then more willing to cooperate by reporting back for follow-up care.

Chapter 11

Organ Transplant Surgery and the Immune Process

Thomas L. Marchioro

I T HAS BEEN known for a long time that grafts of tissue such as skin could be transferred from one site to another on the body of the same individual with indefinite survival of the transplanted structure. Similarly, tissues and organs may be successfully exchanged between identical twins.[1] Such transfers are known as autografts or isografts. When tissues or organs are transplanted between genetically nonidentical individuals of the same species, however, they almost invariably fail, usually within a week or two. These latter type grafts are called homografts or allografts. The factors responsible for this repudiation or rejection of the graft have been shown to be immunologic in nature[2] and are largely determined by the degree of genetic disparity between donor and recipient (histocompatibility)[3] and the immunologic competence of the recipient.[4-6]

Great strides have been made in delineating the histocompatibility antigens in man,[7] and the techniques developed have been used extensively in clinical transplantation of organs from donors to related and unrelated recipients.[8-10] Despite its undoubted value in reducing the severity of the rejection response of the host, however, histocompatibility matching by itself has not ensured indefinite survival of homografts. To obtain prolonged survival, there must also be an alteration in the responsiveness of the host to foreign antigens.

It is generally agreed that the process of rejection is cell mediated, principally by the small, immunologically competent lymphocyte, and is either identical with, or closely allied to, delayed hypersensitivity.[11] Humoral factors are thought to play a subsidiary role, although recent evidence suggests that they may be of greater importance than has been accorded them in the past.[12,13] Whatever the case, it is known that under circumstances in which host immunologic capacity is reduced, homografts may enjoy extended and even indefinite survival.

In man, a variety of pathologic states associated with impaired immunity has been shown to permit long-term survival of test skin grafts. Among these abnormal conditions are hypogammaglobulinemia, Hodgkin's disease, thymic alymphocytosis and other so-called immunologic deficiency diseases.[14]

Note: This work was supported in part by Grants FR-37, AI-08435 and AI-08816 and Contract PH43-67-1435 from the U. S. Public Health Service.

Unfortunately, the individuals suffering from these disorders are rarely in need of an organ homograft, except possibly one which would confer enhanced immunologic potential such as a thymic, bone marrow, or splenic transplant. The usual result of such transfers has been either failure of the graft to take or death from what is believed to be an immunologic reaction against the host by the engrafted immunologically competent cells (graft verus host reaction) .[15] Recently, however, grafts of thymic cells to infants appear to have been successful.[16,17]

Because the usual patient in need of an organ homograft has an intact, normally functioning immune system, interference with its mechanisms is necessary to permit acceptance of the graft. The principal agents presently in clinical use for this purpose are antimetabolites, corticosteroids, antibiotics, x-irradiation, heterologous antilymphocyte antibodies, excision of lymphoid organs such as the thymus, and depletion of lymphocytes by chronic thoracic duct fistula.

IMMUNOSUPPRESSIVE AGENTS

Antimetabolites

Virtually every compound known to affect cellular metabolism has been utilized in the experimental laboratory to achieve immunosuppression. Although many of these have been shown to be effective in inhibiting or preventing both antibody synthesis and the expression of delayed hypersensitivity reactions, only a few have been useful in achieving prolongation of organ homograft survival in the dog, the species most widely used for this purpose. The most successful of these agents has been azathioprine (Imuran®), an imidazole derivative of 6-mercaptopurine. It has become the cornerstone of immunosuppressive regimens in human organ transplantation. Like its parent compound, azathioprine inhibits nucleic acid synthesis. Its pharmacology has been reviewed by Hitchings and Elion.[18] It has been shown to inhibit humoral antibody synthesis, to delay hypersensitivity and homograft rejection, and to possess anti-inflammatory properties. Its principal toxic effects are exerted on the bone marrow, overdosage leading to severe prolonged pancytopenia (Fig. 11-1) .[19]

Corticosteroids

Although initial attempts to prolong organ homograft survival with corticosteroids were generally unrewarding, subsequent work has shown that prednisone in combination with other immunosuppressants has a remarkable adjuvant effect in reversing an established rejection crisis.[19] The exact nature of the immunosuppressive properties of prednisone are unknown. It is known to be a powerful lympholytic agent, is antiphlogistic, an

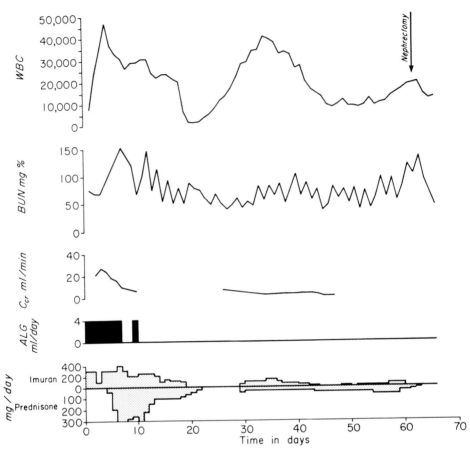

Figure 11-1. Course of a 29-year-old patient who received a renal homograft from his brother. A severe rejection crisis occurred on the fourth posttransplant day, accompanied by fever to 40°C, leukocytosis, oliguria and diminished renal function manifested by a rising blood urea nitrogen (*BUN*) and falling creatinine clearance (*Ccr*). Despite institution of massive doses of prednisone and local graft irradiation (not shown), adequate function did not return. He was dialyzed three times weekly during this interval and, following removal of the nonfunctioning transplant, has been returned to chronic hemodialysis. The graph illustrates the profound depressive effect of relatively modest doses of azathioprine (1 to 2 mg/kg per day) on the peripheral white blood count (*WBC*), particularly in the presence of poor renal function. In this instance, discontinuance of the drug was followed after an interval of several days by complete hematologic recovery.

inhibitor of phagocytosis, stabilizer of lysosomal membranes, and an inhibitor of antibody synthesis.[20] Any one or a combination of these actions could readily explain its effectiveness, as well as its toxicity.

Actinomycin C

This drug, a mixture of three related antibiotics, was isolated from *Streptomyces chrysomallus*. It has been shown to be quite toxic in man[19] and has recently been withdrawn from the market.

X-Irradiation

Whole Body Irradiation

It has been known since the early part of the century that whole body irradiation can suppress antibody formation. Total body irradiation for prolonging graft survival, however, has been disappointing. Although two patients, one in Boston[21] and one in Paris,[22] who were conditioned with total body irradiation are still alive with their original renal homografts still functioning more than ten years later, the majority of patients so treated died early from bone marrow aplasia and sepsis.[23] Today total body irradiation has been abandoned by all centers involved in human organ transplantation. Recent studies in dogs by Epstein and his associates,[24] however, suggest that, under appropriate circumstances, it may still find a place, principally in bone marrow grafting.

Irradiation of the Graft

Local irradiation of the graft has been shown by Hume's group to prolong renal homograft survival in dogs when used as the sole immunosuppressive therapy.[25] They interpret their findings as indicating that local irradiation interferes with both sensitization of the host to the graft and the subsequent destruction mediated by host immunologic defenses. Exactly how this occurs is not known, but the effect has been sufficiently convincing for several groups to have adopted the technique as part of their regimen, using it either prophylactically to prevent a rejection crisis or therapeutically in an effort to reverse an established rejection.[26]

Antilymphocyte Serum and Antilymphocyte Globulin

The constant search for more potent and specific immunosuppressants has led to the development of a variety of new agents, the most promising being antilymphocyte serum or globulin. These antisera are heterologous, being produced by inoculating lymphoid tissue of one species into a different species. Antisera prepared for use in humans have been produced primarily in horses and rabbits. The lymphoid tissues used have been derived from spleen, thymus, lymph nodes, thoracic duct drainage and cultures of human lymphocytes obtained originally from peripheral blood. Inoculating injections have been given intramuscularly, subcutaneously and intravenously with or without adjuvant. Potency of the serum has been assayed by leukoagglutination, lymphocytotoxicity, inhibition of delayed hypersensitivity, transformation of lymphocytes in culture, and inhibition of the mixed lymphocyte culture reaction, among other methods. The antibody has been shown to reside largely in the IgG fraction of serum, although in horses chronically immunized with lymphocytes the antibody tends to concentrate in the equine T fraction with some activity found in the beta globulins. Antilymphocyte plasma, serum and varying degrees of purified globulin derivatives have been prepared and tested. There are many techniques for preparation, assaying, purifying and administering antilymphocyte antibody. The entire subject has been reviewed in several recent publications.[27,28] The mode of action of this substance is not yet clear, although it appears that it acts primarily by destroying the small, long-lived lymphocyte thought to be the principal agent in graft destruction.

Thoracic Duct Drainage

Following the pioneer experiments of McGregor and Gowans[29] interest in thoracic duct drainage as a means of immunosuppression was extended to selected human subjects. It was demonstrated that chronic drainage of lymphocytes from a thoracic duct fistula could interfere with the immune response in man, including depressed antibody formation, inhibition of delayed hypersensitivity and prolongation of homograft survival.[30,31] Some investigators believe it is necessary to drain not only lymphocytes but the lymph as well.[31] Others feel it is necessary to deplete only the lymphocytes.[30]

ORGAN TRANSPLANTS AND IMMUNOSUPPRESSIVE AGENTS

During the past decade, steadily increasing numbers of patients have been treated with organ homografts. The largest experience is with renal transplantation, and this endeavor has been the model for all subsequent undertakings involving transplantation of liver, heart, lung, pancreas, bowel and spleen in man.

Kidney

The most extensive and meaningful statistics on organ transplantation are those applying to the kidney. Since 1963 the National Kidney Transplant Registry, established by the National Academy of Sciences, National Research Council, has been recording human renal transplantation operations. The latest compilation of data supplies information on 1741 kidney transplants performed through January 1, 1968.[32] There were forty-four participating institutions in the United States and forty-five from abroad. Kidneys were obtained from living related volunteer donors in 48 percent of the cases and from human cadavers in 50 percent. Seventy-eight percent of transfers between siblings other than monozygotic or dizygotic twins were functioning at one year, and 76 percent were functioning at two years after transplantation. Comparable figures were reported for parent-to-offspring combinations. When cadaveric organs were used, 45 percent were functioning at one year and 37 percent at two years. The most impressive feature of this report is the remarkable gains made in all categories, particularly with cadaveric organs, when compared to previous reports.

The results just cited were achieved, for the most part, with the so-called "standard" immunosuppressive regimens combining azthioprine, prednisone, actinomycin C and local graft irradiation in varying dosages and schedules. The influence of the newer methods of therapy such as antilymphocytic antibody cannot fairly be assessed from such statistics. Only within the context of the experience at a given institution is it possible to determine with any degree of precision the value of alterations in therapeutic protocol. It becomes necessary therefore to examine individual series in order to appreciate more clearly the influence of the newer agents.

Thymectomy in small animals is associated with a profound immunologic depression. The procedure has been employed to enhance survival of organ transplants in man. Although the initial results obtained by Starzl *et al.* in humans were intriguing,[33] more extensive trials utilizing a random-selection protocol have failed to show any definite advantage enjoyed by thymectomized recipients of renal homografts when compared to those patients who retained this gland.[34]

Similar reasoning was invoked regarding the potential usefulness of splenectomy in potentiating homograft acceptance.[35] Several studies suggested that splenectomy resulted in an obtunded response to a variety of antigens, particularly those introduced intravenously. Several centers have now reported on the effects of splenectomy in human recipients of renal homografts.[36] No clear-cut benefit has been noted. A few groups have found an increased incidence of infectious and thrombotic complications following splenectomy. Recently Pierce and Hume[37] have presented evidence that, although splenectomy is of no value in primary renal homografts, it appears to be useful in preventing rejection of second grafts.

Antilymphocytic antibodies have been used clinically in several centers in the treatment of renal homograft recipients.[9,38,39] The largest series with the longest follow-up is that of Starzl and his associates.[9] Several publications detail their method of preparing, assaying, purifying and testing a variety of antilymphocyte products in animals with the subsequent employment of an ammonium-sulfate-precipitated gamma globulin product in humans.[40,41] It has been used as an adjuvant agent clinically.

Of Starzl's twenty patients, who received homografts from living related volunteer donors $2\frac{1}{2}$ to 3 years ago, 95 percent are still living with the original graft still functioning.[9] The dosages of azathioprine and prednisone necessary to maintain homograft function in these patients have been lower than in any previous group treated at the center, and renal function is as good or better than that previousdy achieved. The administration of antilymphocyte globulin is generally safe, although some nonfatal anaphylactic reactions have occurred.[42] After completion of a course of antilymphocyte globulin, horse protein was found by biopsy in the kidney of only one of twelve patients with renal homografts.[34] Traeger *et al.* have reported similar experiences.[38]

Murray and his associates[43] have used chronic thoracic duct fistula as and aid in treating recipients of renal homografts. Twenty-two patients who received kidneys from blood relatives had a one-year survival of 90 percent when a satisfactorily functioning thoracic duct fistula was added to the basic immunosuppressive regimen of azathioprine and prednisone. In contrast, in a concurrent group of eighteen patients in whom a thoracic duct fistula was either not constructed or failed, the one-year survival rate was only 50 percent. The two groups were otherwise comparable in relation to immunosuppression and histocompatibility.

Kountz and Cohn[44] infuse the standard immunosuppressive drugs directly into the renal artery. With this technique they have performed sixty-one transplants over the past thirty months. Fifty-nine patients are alive and fifty-eight kidneys are functioning.

Other Organs

In the results cited above, the influence of histocompatibility typing was ignored. This is not because of the lack of importance of this factor in determining the outcome, particularly the long-term results, but rather for lack of space. Several recent articles and reviews have been devoted to this subject and are well worth consulting.[7-10]

An additional factor which has not been discussed is the privileged status of the recipient of a renal homograft compared to those individuals receiving hepatic, cardiac, or pulmonary transplants. Following successful insertion of a kidney, diuresis is usually prompt, and blood urea nitrogen and creatinine clearance return rapidly to normal.[19] These easily measured parameters of renal function are sensitive indicators of the functional status of the transplant and as such provide precise moment-to-moment monitors of the rejection process and its control. No such readily available and accurate measures of rejection are available for the heart or lung. With liver grafts, ordinary determinations of liver function appear to be adequate.[45]

Another not inconsiderable advantage of the patient with a renal homograft is the availability of long-term interval support by artificial dialysis if the kidney should fail, either temporarily or permanently. There are no means of substituting for cardiac, pulmonary, or hepatic function for more than a few hours. This has led to replacement of failing liver[45] or cardiac homografts with another homograft. Although these efforts were technically successful, they pose a formidable challenge.

Whether features of the host immune response are peculiar for grafts of different organs is not yet clear. Certain organs tend to evoke lesser responses in certain species, such as liver homografts in pigs. So far, no convincing evidence has been forthcoming that this is the case in man. There is, however, definite evidence from the work of Starzl and associates that human recipients of liver homografts are much more sensitive to the toxic effects of azathioprine on the bone marrow.[45] This appears to be due largely to damage to the xanthine oxidase system which is largely concentrated in the human liver. This same enzyme is widely distributed in the dog, but the same degree of sensitivity of its bone marrow has not been observed, even though the drug is quite hepatotoxic in this species.

From the foregoing considerations, it may be readily appreciated why transplantation of organs other than the kidney has only recently been successful. Despite such problems, however, survival exceeding one year has been achieved in several recipients of liver[45,46] and cardiac homografts.[47,48] Encouraging results have been reported with the pancreas.[49] It appears to be only a matter of time until the same measure of success which now attends cadaveric renal homografts will be achieved with other transplants. Improvements in organ preservation, immunosuppression and histocompatibility typing are being made almost daily. The future for this new therapeutic modality appears extremely promising.

Immunologic Complications of Organ Transplantation

As stated at the outset, to achieve chronic functioning of an organ homograft, it is necessary to induce host acceptance of the transplant by depressing or suppressing the immune mechanisms. Virtually all the agents discussed herein are nonspecific in their actions, resulting in a general lowering of body defenses against any and all foreign antigens including bacteria, fungi, viruses and even neoplasms. These effects have been largely responsible for the fatal and nonfatal complications encountered.

Azathioprine is a potent bone marrow suppressant. As such it can readily produce serious leukopenia (Fig. 11-1). Its dosage must be carefully regulated on a day-to-day basis, by utilizing the peripheral white blood cell count as a guide.[19] In the presence of impaired renal function, its toxicity is increased, and it then becomes necessary to administer reduced dosages of the drug or withhold it altogether. Failure to observe these guidelines has resulted in fatal marrow aplasia. In addition to its increased toxicity in the presence of renal failure, azathioprine also has a heightened effect on the bone marrow in recipients of hepatic homografts.[45]This enhanced toxicity appeared to be especially important in the deaths following early trials of liver transplantation in man.

Prednisone, with its multiple actions, has an even more pronounced effect on the host immune response. It is particularly insidious, the complications for the most part being delayed. In general, infections occurring following prolonged, high doses of this agent are due to a variety of exotic and opportunistic organisms, which are frequently difficult or impossible to eradicate. The patterns of infection observed following renal homotransplantation have been thoroughly reviewed and reported by Rifkind[19,50] and Hill[51] and their associates. A noteworthy feature has been the activation of latent viruses such as the salivary gland virus or that of herpes zoster (Fig. 11-2). In the case of the former agent, widespread cytomegalic inclusion disease has been found at autopsy, usually in conjunction with other agents such as pneumocystis carinii or fungal invaders. Whether the salivary gland virus or pneumocystis carinii is responsible for the so-called transplantation pneumonia[19] or transplant lung[52] is uncertain. There is little doubt, however, among the groups with extensive experience in organ transplantation that the need for prolonged high doses of steroid is responsible for a large fraction of the infectious complications which occur in homograft recipients.

In the past, the complications of infection associated with azathioprine toxicity or with chronic high-dose administration of prednisone have been the principal cause of mortality following homotransplantation of the kidney and liver. With the advent of newer immunosuppressive agents, particularly antilymphocyte globulin, this element of toxicity seems to have been markedly diminished. The utilization of this latter agent during the first four postoperative months has permitted a significant reduction in the amounts of azathioprine and prednisone necessary to maintain graft function, with a resultant decrease in infectious complications. More importantly, infections and other complications can be handled more satisfactorily when they do occur. Starzl *et al.* have been able to control localized hepatic abscesses in recipients of liver homografts for prolonged periods[45]; this complication would have been uniformly and rapidly fatal prior to the introduction of antilymphocyte globulin.

An illustrative case is presented in Figure 11-3. On August 2, 1968, this 10-year-old child received a renal homograft from her father. Immediate

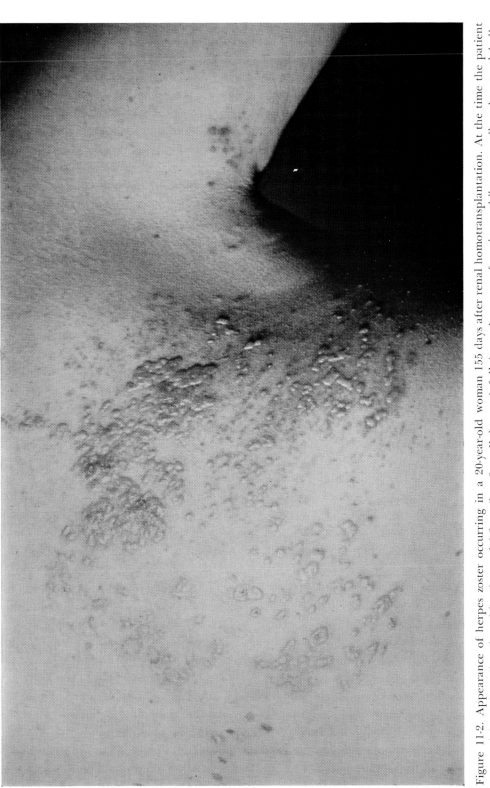

Figure 11-2. Appearance of herpes zoster occurring in a 20-year-old woman 155 days after renal homotransplantation. At the time the patient was receiving 1.0 mg/kg of azathioprine and 0.3 mg/kg of predinisone. A full 4-month course of equine antihuman antilymphocyte globulin had been completed 63 days previously. Other patients have manifested oral herpetic lesions and multiple verrucae vulgaris, principally on the hands and feet.

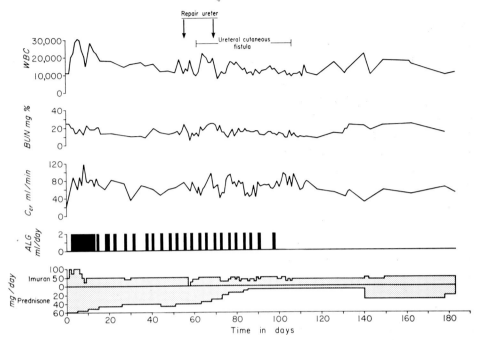

Figure 11-3. Course of a 10-year-old girl who received a renal homograft from her father (see text). *WBC*—peripheral white blood cell count; *BUN*—blood urea nitrogen; *Ccr*—creatinine clearance; *ALG*—equine antihuman antilymphocyte globulin.

function was excellent. At six weeks a ureteral abnormality was noted on an intravenous pyelogram. Exploration revealed necrosis of the distal third of the ureter. Two attempts at ureteroureterostomy failed and a ureterocutaneous fistula persisted. During the next two months it was possible to reduce the prednisone dosage in this girl to 7.5 mg a day and maintain it at this level for several weeks without a change in renal function. Under these circumstances the fistula healed, and the child has remained well with excellent renal function since. Previously, urinary fistulas had been associated with a high mortality.

Although antilymphocyte globulin has been shown to enhance viral and tumor growth in tissue culture and intact animals, there is no evidence to date that similar effects obtain in human subjects, treated with this substance. Its main undesirable side effects are pain, fever, rash and nonfatal anaphylactic reactions. The latter occurred in 15 percent of Starzl's patients. The immunologic and clinical finding in these patients have been reported by Kashiwagi *et al.*[42] Despite these complications, antilymphocyte globulin appears to be relatively safe.

Chronic thoracic duct fistula drainage has been shown to profoundly depress the antibody response to a variety of antigens, including bacteria. Fatal septicemia has been reported in one case.[30] In spite of these potentially lethal effects, Murray *et al.*[43] did not find a significant increase in infections in renal homograft recipients treated with thoracic duct fistula. Further experience with this mode of immunosuppression will be eagerly awaited.

Despite the improvements realized in reducing infectious complications associated with use of immunosuppressive agents in man, the outlook would be bleak if it were not for a remarkable biologic phenomenon which occurs after a variable period of time. First demonstrated by the Woodruffs[53] and termed *adaptation* by them, this phenomenon consists of a reduced tendency on the part of the host to reject the graft immediately. The exact mechanisms, however, remain unclear. An especially enlightening discussion has been given by Murray and colleagues.[54] Whatever the mechanism, the practical consequences of adaptation are that it becomes possible to progressively diminish the stringency of the immunosuppression necessary to maintain the homograft, usually after the first few months. This reduction in drug dosage permits more normal responses to environmental bacteria and allows a wide variety of medical and surgical problems to be dealt with successfully. Penn and his associates[55] have recently reviewed their experience with surgical complications arising in transplant recipients, including acute appendicitis, upper gastrointestinal bleeding and intra-abdominal abscesses. Many of these complications were successfully treated, particularly those that occurred after the first four months.

Finally, a word must be said about the possible development of cancer in the chronically immunosuppressed patient. There is evidence in both animals and man that the same agents that permit chronic homograft survival in human subjects can also potentiate the implantation and subsequent growth of viable malignancies.[56] It is not yet clear that these agents will increase the number of spontaneously arising neoplasms. There is, however, experimental evidence to suggest that reticuloendothelial malignancy may arise in the experimental animal under chronic immunologic stress.[57] In this connection, the recent report[58] of the development of lymphoid neoplasms in five recipients of renal homografts is of considerable interest. To date the incidence of *de novo* development of cancer is low and probably does not exceed the expected occurrence in the general population. More information must be gathered, however, for a more definitive opinion.

REFERENCES

1. Murray, J.E., Merrill, J.P., and Harrison, J.H.: Renal homotransplantation in identical twins. *Surg Forum,* 6:432, 1955.
2. Medawar, P.B.: Immunology of transplantation. *Harvey Lect,* 52:144, 1958.

3. Cepellini, R.: The genetic basis of transplantation. In Rapaport, F.T., and Dausset, J. (Eds.) : *Human Transplantation.* New York, Grune & Stratton, 1968, pp. 21-34.

4. Dammin, G.J., Couch, N.P., and Murray, J.E.: Prolonged survival of skin homografts in uremic patients. *Ann NY Acad Sci, 64:*967, 1957.

5. Good, R.A., Varco, R.L., Aust, J.B., and Zak, S.J.: Transplantation studies in patients with agammaglobulinemia. *Ann NY Acad Sci, 64:*882, 1957.

6. Rosen, F.S., Gitlin, D., and Janeway, C.A.: Alymphocytosis, agammaglobulinemia, homografts and delayed hypersensitivity: Study of case. *Lancet,* 2:380, 1962.

7. Curtoni, E.S., Mattiuz, P.L., and Tosi R.M. (Eds.) : *Histocompatibility Testing.* Copenhagen, Munksgaard, 1967.

8. Lee, H.M., Hune, D.M., Vredovoe, D.L., Mickey, M.R., and Terasaki, P.I.: Serotyping for homotransplantation. IX. Evaluation of leukocyte antigen matching with the clinical course and rejection types. *Transplantation,* 5:1040, 1967.

9. Starzl, T.E. *et al.:* Heterologous antilymphocyte globulin, histocompatibility matching, and human renal homotransplantation. *Surg. Gynec Obstet, 126:*1023, 1968.

10. Terasaki, P.I., Vredevoe, D.L., and Mickey, M.R.: Serotyping for homotransplantation. X. Survival of 196 grafted kidneys subsequent to typing. *Transplantation,* 5:1057, 1967.

11. Gowans, J.L.: The role of lymphocytes in the destruction of homografts. *Brit Med Bull, 21:*106, 1965.

12. Starzl, T.E. *et al.:* Schwartzman reaction after human renal homotransplantation. *New Eng J Med, 278:*642, 1968.

13. Williams, G.M. *et al.:* "Hyperacute" renal homograft rejection in man. *New Eng J Med, 279:*611, 1968.

14. Good, R.A. (Ed) : *Immunologic Deficiency Diseases in Man.* New York, The National Foundation, 1967.

15. Hathaway, W.E. *et al.:* Graft-vs-host reaction following a single blood transfusion, *JAMA, 201:*1015, 1967.

16. August, C.S. *et al.:* Implantation of a foetal thymus, restoring immunological competence in a patient with thymic aplasia (DiGeorge's syndrome) . *Lancet, 2:* 1210, 1968.

17. Cleveland, W.W., Fogel, B.J., Brown, W.T., and Kay, H.E.M.: Foetal thymic transplant in a case of DiGeorge's syndrome. *Lancet,* 2:1211, 1968.

18. Hitchings, G.H., and Elion, G.B.: Chemical suppression of the immune response. *Pharmacol Rev, 15:*365, 1963.

19. Starzl, T.E.: *Experience in Renal Transplantation.* Philadelphia, W. B. Saunders Co., 1964.

20. Mannick, J.A., and Egdahl, R.H.: Endocrinologic agents. In Rapaport, F.T. and Dausset, J. (Eds.) : *Human Transplantation.* New York, Grune & Stratton, 1968, pp. 472-481.

21. Merrill, J.P. *et al.:* Successful homotransplantation of kidney between nonidentical twins. *New Eng J Med, 262:*125, 1960.

22. Hamburger, J. *et al.:* Transplantation d'un rein entre jumeaux non monozygotes apres irradiation du receveur. Bon Fonctinnement au quatrieme mois. *Presse Med, 67:*171, 1959.

23. Murray, J.E. *et al.:* Kidney transplantation in modified recipients. *Ann Surg, 156:* 337, 1962.

170   *Surgery and the Allergic Patient*

24. Epstein, R.B., Storb, R., Ragde, H., and Thomas, E.D.: Cytotoxic typing antisera for marrow grafting in lettermate dogs. *Transplantation, 6:*45, 1968.
25. Wolf, J.S., McGavic, J.C., and Hume, D.M.: Inhibition of the effector mechanism of transplant immunity by local graft irradiation. *Surg Gynec Obstet, 128:*584, 1969.
26. Murray, J.E., Barnes, B.A., and Atkinson, J.: Fifth report of the human kidney transplant registry. *Transplantation, 5:*752, 1967.
27. Medawar, P.B.: Biological effects of heterologous antilymphocyte sera. In Rapaport, F.T., and Dausset, J. (Eds.): *Human Transplantation.* New York, Grune & Stratton, 1968, pp. 501-509.
28. Wolstenholme, G.E.W. (Ed.): *Antilymphocyte Serum.* London, J. & A. Churchill Ltd, 1967.
29. McGregor, D.D., and Gowans, J.L.: Antibody response of rats depleted of lymphocytes by chronic drainage from the thoracic duct. *J Exp Med, 117:*303, 1963.
30. Graber, C.D., Fitts, T., Williams, A.V., Artz, C.P., and Hargest, T.S.: Immunologic competence in the lymph-dialyzed patient. *Surg Gynec Obstet, 128:*1, 1969.
31. Irvin, G.L., and Carbone, P.P.: Immunosuppression with lymph depletion in man. *Surg Gynec Obstet, 124:*1283, 1967.
32. Sixth report of the human kidney transplant registry. *Transplantation, 6:*944, 1968.
33. Starzl, T.E. et al.: Chronic survival after human renal homotransplantation; lymphocyte-antigen matching, pathology and influence of thymectomy. *Ann Surg, 162:*749, 1965.
34. Starzl, T.E., Brettschneider, L., Martin, A.J., Groth, C.G., Blanchard, H., Smith, G.V., and Penn, I.: Organ transplantation, past and present. *Surg Clin N Amer, 48:*817, 1968.
35. Starzl, T.E., Marchioro, T.L., Talmage, D.W., and Waddell, W.R.: Splenectomy and thymectomy in human renal transplantation. *Proc Soc Exp Biol Med, 113:*929, 1963.
36. Gleason, R.E., and Murray, J.E.: Report from kidney transplant registry; analysis of variables in the function of human kidney transplants. I. Blood group compatibility and splenectomy. *Transplantation, 5:*343, 1967.
37. Pierce, J.C., and Hume, D.M.: The effect of splenectomy on the survival of first and second renal homotransplants in man. *Surg Gynec Obstet, 127:*1300, 1968.
38. Traeger, J. et al.: Studies of antilymphocyte globulins made from thoracic duct lymphocytes. *Transplantation Proceedings, 1:*445, 1969.
39. Shorter, R.G. et al.: Antilymphoid sera in renal allotransplantation. *Arch Surg, 97:*323, 1968.
40. Iwasaki, Y. et al.: The preparation and testing of horse antidog and antihuman antilymphoid plasma or serum and its protein fractions. *Surg Gynec Obstet, 124:*1, 1967.
41. Kashiwagi, N. et al.: Improvements in the preparation of heterologous antilymphocyte globulin with special reference to absorption and diethylaminoethyl cellulose batch production. *Surgery, 65:*477, 1969.
42. Kashiwagi, N., Brantigan, C.O., Brettschneider, L., Groth, C.G., and Starzl, T.E.: Clinical reactions and serologic changes following administration of heterologous antilymphocyte globulin to human recipients of renal homografts. *Ann Intern Med, 68:*275, 1968.
43. Murray, J.E. et al.: Five years experience in renal transplantation with immunosuppressive drugs: Survival, function, complications and the role of lymphocyte depletion by thoracic duct fistula. *Ann Surg, 168:*416, 1968.

44. Kountz, S.L., and Cohn, R.: Initial treatment of renal allografts with large intra-renal doses of immunosuppressive drugs. *Lancet, 1:*338, 1969.

45. Starzl, T.E. *et al.:* Orthotopic homotransplantation of the human liver. *Ann Surg, 168:*392, 1968.

46. Starzl, T.E. *et al.:* Clinical and pathologic observations after orthotopic transplantation of the human liver. *Surg Gynec Obstet, 128:*327, 1969.

47. Barnard, C.N.: What we have learned about heart transplants. *J Thorac Cardiovasc Surg, 56:*457, 1968.

48. Cooley, D.A., Bloodwell, R.D., and Hallman, G.L.: Cardiac transplantation for advanced acquired heart disease. *J Cardovasc Surg, 9:*403, 1968.

49. Lillehei, R.C. *et al.:* Transplantation of the intestine and pancreas. *Transplantation Proceedings, 1:*230, 1969.

50. Rifkind, D., Marchioro, T.L., Schneck, S., and Hill, Jr., R.B.: Systemic fungal infections complicating renal transplantation and immunosuppressive therapy—clinical, microbiologic, neurologic and pathologic features. *Amer J Med, 43:*28, 1967.

51. Hill, R.B., Dahrling, B.E., Starzl, T.E., and Rifkind, D.: Death after transplantation, an analysis of 60 cases. *Amer J Med, 42:*237, 1967.

52. Slapak, M., Lee, H.M., and Hume, D.M.: "Transplant lung" and lung complications in renal transplantation. In Dausset, J., Hamburger, J., and Mathe, G. (Eds.) : *Advance in Transplantation.* Copenhagen, Munksgaard, 1967.

53. Woodruff, M.F.A., and Woodruff, H.G.: The transplantation of tissue; auto- and homotransplants of thyroid and spleen in the anterior chamber of the eye, and subcutaneously, in guinea pigs. *Philos Trans Roy Soc (London), 234:*559, 1950.

54. Murray, J.E. *et al.:* Analysis of mechanism of immunosuppressive drugs in renal homotransplantation. *Ann Surg, 160:*449, 1964.

55. Penn, I. *et al.:* Surgically correctable intra-abdominal complications before and after renal homotransplantation. *Ann Surg, 168:*865, 1968.

56. Schwartz, T.S.: Are immunosuppressive anticancer drugs self-defeating? *Cancer Res, 28:*1452, 1968.

57. Schwartz, R., Andre-Schwartz, J., Armstrong, M.Y.K., and Beldotti, L.: Neoplastic sequelae of allogenic diseases. I. Theoretical considerations and experimental design. *Ann NY Acad Sci, 129:*804, 1966.

58. Penn, I., Hammond, W., Brettschneider, L., and Starzl, T.E.: Malignant lymphomas in transplantation patients. *Transplantation Proceedings, 1:*106, 1969.

Appendix

Contact Dermatitis from Chemicals to Which Surgical Patients Are Exposed

Leonard W. Maxey
President, Hollister-Stier Laboratories
Spokane, Washington

M ORE AND MORE TODAY people are exposed to an increasing number of chemicals in their environment. The surgeon and his patient are not immune to the growing number of instances of possible contact with adhesives, anesthetics, antibiotics, drugs, germicidal agents, metals, plastics, rubber gloves, soaps and detergents. As a result, contact dermatitis from chemicals is recognized as an increasingly important chemical problem.

It should be noted here that where the causative agent in dermatitis is a chemical there is no desensitization treatment available. To assist the physician in determining causative chemical agents, commercial laboratories* have made available patch test material and information literature.

Patch test screening is recognized as a useful means of determining chemical sensitizers responsible for contact dermatitis. They are supplied in individual syringe or applicator vials depending on the vehicle required. A selection of chem tests based on surgical contact would include the following available chem tests:

1. Caine Mix (Benzocaine, Nupercaine, Surfacaine)
2. Medimix (Furacin, Histadyl, Vioform)
3. Paraben Mix (Ethyl, Methyl and Propyl parabens)
4. Rubber Mix (Mercaptobenzothiazole, Tetramethythiuram Disulfide)
5. Formalin
6. Mercuric Chloride, ammoniated
7. Merthiolate, sodium
8. Neomycin
9. Balsam of Peru
10. Coal Tar
11. Bithionol
12. Hexachlorophene

*Hollister-Stier Laboratories, Spokane, Washington.

13. Polybrominated Salicylanilide
14. Resorcinol
15. Nickel Sulfate

Patches should be placed on grossly normal nonhairy skin. Patch testing WILL NOT be reliable if performed on skin that is in any way inflamed or dermatitic The upper back and the upper outer arm are the most satisfactory sites for patch testing. Areas such as the inner arm, abdomen or anterior thigh are definitely less sensitive than either the upper outer arm or upper back. When only a few pach tests are being performed, the upper outer arm is probably the location of choice. When more than four or five tests are indicated, the upper back is more suitable.

In general, if a test substance is suspected and might cause a severe reaction on testing, then that substance should be applied separately and away from the main body of patch tests. Patch testing should not be undertaken on patients taking systemic corticosteroids.

Cleanse the site with alcohol. The chemicals to be tested are placed on the adhesive patch test covers available from Hollister-Stier Laboratories according to directions supplied.

The patch test reaction depends on sensitivity of the patient, concentration of the substance, and length of time the substance is in contact with the skin.

1. *Maintain Patch Contact 48 hours.*

   However the patient should be instructed to remove, at once, any patch which itches or burns severely.

2. *Reading Should Be Made After an Interval of at Least 30 Minutes From Time of Removal of the Patch.*

   A longer time interval is desirable. Time may be saved by having the patient remove the tapes from two to four hours before the scheduled return visit. This time interval allows the skin to recover from mild erthema or dermagraphia due to the pressure of the patch. The true erythema of specific reaction will persist for hours or possibly days.

To aid identification of patch test sites, it is helpful to mark them at the time of initial application. The skin marking pencil available with the Hollister-Stier patch tests has proved to be very helpful. The markings should last for the 48-hour period without smudging and can be removed with alcohol.

3. *Doubful 1+ Reactions Can Be Observed at 24-Hour Intervals Following the Original Exposure.*

   Erythema that persists or increases is probably significant as an allergic reaction.

4. *A Significant Reaction to Patch Tests Reproduces the Eczematous Process.*

5. *Delayed Patch Test Readings Should Always Be Carried Out Even if the Initial Reading Is Negative.*

   Certain allergens such as Neomycin and Formalin frequently give delayed positives and could easily be overlooked were it not for later readings. It is desirable to do a delayed reading as 48 hours and again four or five days after the removal of the patches. A markedly delayed patch test reaction may indicate sensitization by the test material.

6. *Often But Not Invariably a Positive Patch Test Site Itches.*

7. *When a Single Bulla Occurs, it Usually Denotes a Primary Irritant. There May Be Exceptions to This.*

   The concentrations of chemicals found in the Hollister-Stier standard patch tests are designed to minimize the likelihood of irritant reactions. However irritant skin reactions may still occur in individuals with overly sensitive skin. These irritant reactions, however, will be mild.

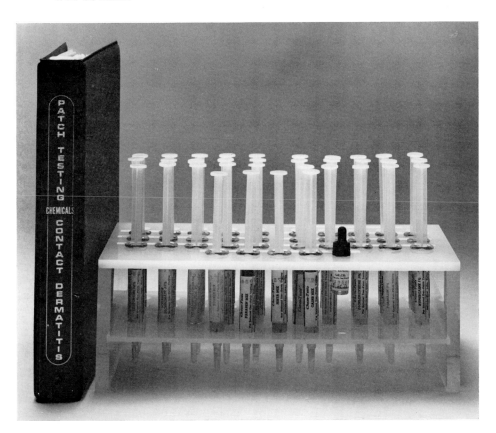

Primary Irritant

    A. Primary irritant may cause a reaction in a few minutes or within an hour.

    B. A primary irritant reaction tends to fade rapidly after patch is removed.

    C. A primary irritant reaction tends to stay within the perimeter of the patch. An allergic reaction tends to spread beyond the confines of the patch.

The pseudo or false positive skin reaction is one of the problems of interpreting patch test responses. The above points are merely guidelines for interpreting the irritant reaction. Often it is necessary to repeat tests when a patient's skin is less irritable. This second test may indicate that this reaction is allergic or an irritant. Artifacts from pressure may occur when solid substances such as pieces of plastic are used for testing. Such pressure artifacts do not occur with the materials included in the Hollister-Stier standard test kit. The following precautions should be observed in patch testing:

    1. Avoid primary irritants unless they are adequately diluted.

    2. Avoid patch tests in an acute or widespread dermatitis.

    3. Avoid testing repeatedly with agents known to sensitize easily.

    4. Do not let test material remain on the skin too long (never longer than five days).

Index

177

R

Rabies vaccine, 30
Racephedrine-containing products, 29
Radiopaque iodine, 21
　and bronchography, 90
　see also Contrast media
Rake retractors, 87
Rales, 38
Rash, clearing, 18
Reactions
　allergic, 13-14
　to protein material, 133
　to transfusions, 133
Reactions to drugs, 4
　of allergic origin, 131
　epinephrine and, 5
　skin tests for, 20
　surgical patient and, 116, 134
"Rebound"
　bronchospasm, 42
　congestion, 18
Recalcitrant asthma, 40
Recording drug sensitivity, 3
Rectal
　bleeding, 110
　suppository, 17
　　aminophylline, 40
Relaxin, 5
Renal
　function, 42
　transplantation, 162-163
　see also Kidneys
Residual volume, measurement of, 69
Respirators
　Berg, 88
　Engstrom unit, 88
　hand-operated, 7
Respiratory
　allergy and horse serum, 21
　allergy and inhalant allergens, 26
　cycle, study, 70
　failure, 94, 98-104
　infections, 14
　　and asthma, 16
Restlessness, 40
Resuscitation
　cardiac, external, technique, 9
　pulmonary, 10 (Fig.)
Reticuloendothelial malignancy, 168
Rhinitis, 13
　children and, 32, 46
　and drug reactions, 116
　and ear, nose and throat disease, 151
　and nose operation, 23, 24, 151

　perennial and intolerance to aspirin, 5
　and preoperative care of patients, 18
Rhonchi, 38
Rib spaces, increased, 65
Right middle lobe (RML), 106
Ring distortion, 103
Ring punch excision, 157
RML disease in children, 106
Roentgenograms and bronchography, 91
Rubber dermititis, 140

S

Salicylates, 5
Salivary gland veins, 166
Salivation, increase in, 120
Salt, loss of, 42
Sanborn X-Y Recorder, 70
Scratch test, 5
　and local anesthetics, 20
　and tetanus antiserum, 29
Screening children for allergic traits, 25
Seasonal allergy, 16
　and drug reactions, 116
Seasonal bronchial asthma, 14
Secobarbital, 20
Secretions and tracheostomy, 87
Sedation and children, 28
Sedatives, 28
　therapy, 136
　use in asthmatic patients, 41
Self-medication and asthma, 17
Sensitivity
　–producing drugs, 3
　tests of sputum cultures, 17
　tetanus antitoxin, 21
Sensitization
　basic, 13
　horse serum, 21
　reaction, 118
　unnecessary, 4
Septal deviation, 24, 154
Serotonin, 124
Serum
　antilymphocyte, 161
　bovine, 21
　electrolyte concentration, 42, 43
　horse, 5, 21
　sickness and use of procaine, 20
　　treatment, 132
Shock
　adrenal insufficiency and, 23
　and anaphylaxis, 3
　and hydrocortisone, soluble, 19
Side effects to antibiotics, 28